THE UNIVERSITY OF NORTH CAROLINA

SOCIAL STUDY SERIES

HUMAN FACTORS IN COTTON CULTURE

THE UNIVERSITY OF NORTH CAROLINA

SOCIAL STUDY SERIES

Odum and Johnson: *The Negro and His Songs*..........$3.00
Puckett: *Folk Beliefs of the Southern Negro*............ 5.00
Odum and Johnson: *Negro Workaday Songs*............ 3.00
Odum and others: *Southern Pioneers*.................. 2.00
Pound: *Law and Morals*.............................. 2.00
Giddings: *The Scientific Study of Human Society*........ 2.00
Odum and Willard: *Systems of Public Welfare*.......... 2.00
Branson: *Farm Life Abroad*.......................... 2.00
Ross: *Roads to Social Peace*......................... 1.50
Willey: *The Country Newspaper*...................... 1.50
Jordan: *Children's Interests in Reading*............... 1.50
Odum: *Public Welfare and Social Work*................ 1.50
North: *Social Differentiation*........................ 2.50
Knight: *Among the Danes*........................... 2.50
Steiner and Brown: *The North Carolina Chain Gang*.... 2.00
Lou: *Juvenile Courts in the United States*............. 3.00
Carter: *The Social Theories of L. T. Hobhouse*.......... 1.50
Brown: *A State Movement in Railroad Development*.... 5.00
Miller: *Town and Country*........................... 2.00
Mitchell: *William Gregg: Factory Master of the Old South* 3.00
Metfessel: *Phonophotography in Folk Music*............ 3.00
Wager: *County Government in North Carolina*.......... 5.00
Brown: *Public Poor Relief in North Carolina*............ 2.00
Walker: *Social Work and the Training of Social Workers* 2.00
Herring: *Welfare Work in Mill Villages*................ 5.00
Vance: *Human Factors in Cotton Culture*.............. 3.00

The University of North Carolina Press
Chapel Hill, N. C.

The Baker and Taylor Co.
New York

Oxford University Press
London

Maruzen-Kabushiki-Kaisha
Tokyo

HUMAN FACTORS IN COTTON CULTURE

A Study in the Social Geography of the American South

BY

RUPERT B. VANCE, PH. D.

Research Associate, Institute for Research in Social Science, University of North Carolina

CHAPEL HILL

THE UNIVERSITY OF NORTH CAROLINA PRESS

1929

THIS BOOK IS

FOR

MY FATHER AND MY MOTHER

What a royal plant it is! The world waits in attendance on its growth; the shower that falls whispering on its leaves is heard around the earth; the sun that shines on it is tempered by the prayers of all the people; the frost that chills it and the dews that descend from the stars are noted. . . . Its fiber is current in every bank, and when loosing its fleeces to the sun it floats a snowy banner that glorifies the fields of the humblest farmer, that man is marshalled under a flag that will compel the allegiance of the world and wring a subsidy from every nation on earth. —Henry W. Grady.

Cotton is the master of them all, in spring as well as autumn, in winter as in summer. Yellow or white or black, all men in the South are slaves of cotton, subject to its power, prospering as those white fields flourish, and failing as they fail. —Dorothy Scarborough.

PREFACE

THE SETTLEMENT of America, viewed by the social geographer, involved an adjustment of previously existent cultures to such factors of natural environment as topography, climate, and societies of plants and animals. The South early in its development was conditioned by the demands of the cotton plant. "Whenever man depends upon agriculture and has found a permanent abode," writes R. Mukerjee, "the growing of different staple crops such as rice, wheat, or Indian corn, and the rearing of different domestic animals, selected from among the native stock of a region, govern not merely man's interests and habits, but also his social organization."

"In the Belt—Black, Cotton, or Bible, as you prefer," a flippant journalist has put it, "cotton is Religion, Politics, Law, Economics, and Art." Without accepting geographic determinism, one must admit that much that is distinctive of southern culture, its plantation system, its sectionalism, its agricultural life, its rural practices, has developed as a kind of complex around the cotton plant. Without pushing this thesis to any unreasonable limits, the present study has grown out of an attempt to estimate the status of the human factors in cotton culture. The warmth of an emotional interest in the South has as far as possible been restrained by an appeal to the cold and impartial fact. It must be admitted, however, that the great human nexus surrounding cotton culture is too intricate to be set forth adequately by statistics and cases. The primeval, unconquered data of society always prove greater than man's conventionalized representations. Such a study, then, is likely to lead to a sense of failure—to a feeling that the cotton culture complex is there, but that it has not been put on paper. Nevertheless, though the picture be incomplete, the writer hopes that it is not misleading.

This volume is planned as a part of a series of studies in the regional South undertaken by the Institute for Research in Social Science at the University of North Carolina. Taking for their general subject the southern regional field, they attempt to set forth the culture peculiar to the American South in terms of its conditioning by natural environment. The writer, as his part of the task, hopes to complete within the next three years a human geography of the South.

It is a pleasure and a duty to acknowledge my indebtedness to the experts of the United States Department of Agriculture. Such men as C. O. Brannen, E. L. Kirkpatrick, J. T. Sanders, B. R. Coad, W. J. Spillman, O. E. Baker, and Dr. Joseph Goldberger, until his recent death a member of the U. S. Public Health Service, have no doubt become accustomed to anonymity by seeing their researches quoted as Government Bulletin No. 2. It is to such caliber of scholarship as theirs that the publications of the United States government owe their high scientific rank. No less emphatic are my obligations to the historians of the South, W. E. Dodd, M. B. Hammond, Ulrich B. Phillips, the late John Spencer Bassett, and Walter L. Fleming. Their researches must become the common property of all who seek to understand the American South.

Acknowledgments are due Dr. Howard W. Odum, Director of the Institute for Research in Social Science and

Kenan Professor of Sociology, for direction, encouragement, and aid in securing case studies of the farmer and his cotton. Dr. E. C. Branson, Kenan Professor of Rural Social Economics, and Miss H. R. Smedes, librarian, aided in the use of the immense amount of materials on southern agriculture and rural life in the files of the Rural Social Economics Library. Miss Katharine Jocher, Assistant Director of the Institute for Research in Social Science, made valuable suggestions, and Miss Nancy Herndon prepared the manuscript. The editorial staff of the University of North Carolina Press has been very kind to an author whose ignorance of form, appearance, and technology has proved amazing even to himself. On the other hand, the writer is compelled to reserve for himself the undivided distinction of having made every mistake to be found in this volume.

R. B. V.

CHAPEL HILL, N. C.
January 19, 1929

CONTENTS

CHAPTER I

COTTON AND REGIONALISM

THE REGIONS OF THE UNITED STATES

THAT THE extraordinary diversity of geographic conditions in the United States is accompanied by a surprising uniformity of economic opportunity and social conditions and a similarity of culture has often been commented upon. Russia, for instance, is the only nation which compares with the United States in the variety and extent of agricultural regions. Dr. Oliver E. Baker[1] has traced our unity of culture amid diversity of environments to three factors: (1) the strong pressure of economic competition, (2) the facilities for the dissemination of new ideas, (3) and the mobility of capital and labor. "I suppose," writes Professor E. A. Ross, "no large population shows so faint and doubtful a response to region as we Americans. Never before were folk of forest or valley, of sea or river delta, so little insulated. Our education, reading matter, films, sports, standardized articles of consumption, religious denominations, trade and professional unions, political parties and common institutions pull us into a national or at least a sectional plane."[2] Frederick Jackson Turner sees in the cultural homogeneity of the United States a lack of the

[1] See his "Agricultural Regions of North America," Part I, in *Economic Geography* (Oct., 1926), pp. 459–93.

[2] In his Introduction to *Regional Sociology*, by R. Mukerjee, p. ix.

historical attitudes of Europe. "There are not in the United States," he writes, "the historic memories of so many national wrongs and wars. . . . There is not here the variety of languages nor races nor the sharp contrasts in cultural types; there has not been the same bitterness of class conflicts; nor the same pressure of economic need, inducing the various regions to seek by arms to acquire the means of subsistence, the control of natural resources."[3]

It is in the traits of her urban-industrial culture that America is most homogeneous; in her rural life she displays more diversity. Possibly the most satisfactory classification of the natural regions of the United States is that worked out by Dr. Oliver E. Baker.[4] To him an agricultural region is a large area of land characterized by a homogeneity in crops grown but sufficiently unlike adjacent areas to be noticeable. The economic culture of certain plants and consequently the agricultural regions are dependent upon the following physical conditions:

1. Moisture conditions, rainfall, and rate of evaporation.
2. Temperature conditions, length of growing season.
3. Topography, contour of the land.
4. Soils, physical, chemical, and bacteriological factors.[5]

In regard to climatic factors the United States falls roughly into four areas: a cold northern, a warm southern, a moist eastern, and a dry western region.[6] In regard to soils there are three main regions: the central plains, largely dark-soiled grasslands; the East and South, largely light-colored forest lands; and the western arid

[3] "Sections and Nations," *The Yale Review* (Oct., 1922), p. 2.
[4] *Op. cit.*, p. 468. [5] *Ibid.*, p. 460. [6] *Ibid.*, p. 467.

lands.[7] In addition to physical factors, W. J. Spillman[8] sees biological and economic factors as determining plant regions. Bacterial content of soils, the preference of alfalfa for alkaline soils, and the relation of soils and climates to insect pests and fungous diseases he classifies as biological. Under economic factors determining plant production are the value per unit of weight, and the distance to market.

The main plants affected by these various factors are corn, wheat, cotton, oats, and hay. Together they normally occupy more than 30,000,000 acres apiece on American farms, taking more than 87 per cent of the total crop area of the country.[9] In combination with live stock they make up the various types of farming. Beginning with the South, Dr. Baker lists the following agricultural regions of the eastern part of the United States:

1. Subtropical Crops Belt.
2. Cotton Belt.
3. Middle Atlantic Trucking Region.
4. Corn and Winter Wheat Belt.
5. Corn Belt.
6. Hay and Dairying Belt.
7. Spring Wheat Area.[10]

The western United States he divides according to crops produced into the following areas:

8. Grazing and Irrigated Crops Region.
9. Columbia Plateau Region.
10. Pacific Subtropic Crops Region.
11. North Pacific Hay, Pasture, and Forest Region.

---

7 *Ibid.*, p. 466.

8 In *Distribution of Types of Farming in the United States*, Farmers' Bulletin 1289, p. 3.

9 *Ibid.*, p. 2.

10 Dept. of Agriculture *Yearbook*, 1921, Fig. 2, p. 416; *ibid.*, p. 472.

It is useless to deny that out of these different regions with the different products come varying patterns of rural life. Economic backgrounds, habits of life, and cultural routine form significant contrasts.[11] The Kansas wheat farmer uses casual labor and tills broad and productive acres with the most modern of farm machinery. The hog and beef producer of the Corn Belt, slave to his plow, drill, and cultivator until the corn is laid by, rushed for a spell at harvest, spends most of his year in pretty leisurely fashion, watching his live stock turn corn and roughage into good sound meat. The wheat farmer of the Great Plains, the apple producer of the Northwest, the orange grower, and the raisin producer each spends one period of the year in throbbing effort followed by long stretches with nothing special to do. The poultryman, the dairyman and butter fat producer, faced with a steady round of daily duties, fewer lulls, and fewer periods of intensity, has a life and a diet both full and varied with something new for almost every day in the year. The truck farmer of the South Atlantic coast, a hand and knee farmer on small patches, keeps in close touch with a sophisticated urban market and suffers no lack of vitamins in his diet. The southern cotton tenant, with his one mule and single plow, living on salt pork and corn bread, has cotton to plant, chop, and pick, and nothing to do until next year. Provided due consideration is given to America's common institutions and her ubiquitous industrial culture with its standardized products, it may be helpful to regard these natural regions in terms of the culture areas of the anthropologist.

---

[11] See Mordecai Ezekiel in *Farm Income and Farm Life,* ed. Sanderson, pp. 82–83. The present writer, unable to improve upon many of Mr. Ezekiel's apt phrases, has adopted them outright.

Each of these regions has its own resources and economic capacities based on its products and partly determined, as Professor Turner has said, when the geological foundations were laid down. On topography and agricultural production have been erected industrial superstructures which have created other economic interests in fabricated goods. It is these rival economic interests that have resulted in sectionalism. A section may be defined as a region seeking to realize economic interests in political action. It is regionalism in politics that has given content to the designation of the American Congress as "an assembly of geographic envoys." It is Professor Turner again who has given this view its brilliant expression: "We in America are in reality a federation of sections rather than states. State sovereignity was never influential except as a constitutional shield of the section. In political matters the states . . . act in sections and are responsible to the respective interests and ideals of these sections. Party policy and congressional legislation emerge from a process of sectional contests and sectional bargainings. . . . Legislation is the result of sectional adjustments to meet national needs."[12] The tariff for the industrial North Atlantic states and the agricultural relief demanded by the Farmers' Block of the Corn, Wheat, and Cotton Belts offer an obvious contemporary illustration.

It has been suggested that the reason for America's cultural unity is the absence of national and racial differences. Again this is an overstatement. Outside cities, there exist racial groups definitely related to regions and to production of particular types of crops. Variety is

[12] *Op. cit.*, pp. 6–7.

added to American economic and cultural life by occasional settlements of Italians, French, and Germans engaged in vine and truck culture, Mexican casuals in Texas, Japanese and Chinese farmers in California, and Poles in the Connecticut Valley. The place of the Pennsylvania Dutch and of the Scandinavians and Teutons in the wheat culture of the Central Plains is more important. New York Yankees, Chester County Quakers, Georgia Crackers, though of the same racial stock have developed differently. As Ezekiel points out each group has "its own standards evolved out of custom, tradition, and environment."[13]

The Census of 1920 revealed interesting data on the economic levels of the different regions.[14] Over one-third of the value of farm property in the United States and two-fifths of the value of the farm land were found in the Corn Belt. The Corn Belt also contained one-fourth of the value of live stock in the country, and about one-half of the value of farm implements and machinery was reported from the Corn Belt and the Hay and Pasture Region. The value of farm buildings was found to be highest in southern Pennsylvania, next in the Corn Belt, the Spring Wheat Area, and the Hay and Dairying Region. It was lowest in the Cotton Belt. Tractors were located chiefly in the Corn Belt, Spring Wheat Area, and South Pacific Region. Over one-third of the automobiles on farms were found in the Corn Belt, and the rural telephones were concentrated in the Corn Belt and the Hay and Dairying Region. The lowest ratios for automobiles and telephones were found in the Cotton Belt. Running

---

[13] Sanderson, *op. cit.*, p. 83.

[14] See "Graphic Summary of American Agriculture" in Dept. of Agriculture *Yearbook*, 1921, pp. 493–506.

water and electric lighting systems in farm homes were found mostly in the Hay and Dairying regions of New England and the South Pacific Area. When urban centers are included it is shown that the North Atlantic Coast states, with less than a third of the population, paid over one-half of the income and profit taxes appropriated for the nation as a whole.[15] A study of reading habits of the population shows that New England, New York, and New Jersey have distinct preëminence. The draft records reveal the fact that the best physical fitness was found in North Central and Mountain states, and the lowest in the industrial Northeast. A mapping of the men and women of eminence as listed by *Who's Who* shows that over half of those included live in the northeast Atlantic states, although a little less than half were from that section.

THE SOUTH AND COTTON

The inevitable illustration of the trends discussed above is the South. It furnishes the region most definitely committed to the production of one economic plant, the staple of cotton. In the Cotton Kingdom it has furnished the most sharply defined section and has led the most aggressive sectional movement in the history of our country. In its Negroes it possesses the largest and most clearly defined racial group in the United States. This group is historically associated with cotton culture, and the association continues. The South is relatively among the least urban and industrial of the regions of the United States. This, of course, is another way of saying

[15] See also Turner, "Sections and Nations," *The Yale Review* (Oct., 1922), pp. 11–12.

that its habits of rural life make up a comparatively large part of its culture. Its leading agricultural product, cotton, may also be expected to play a larger part in the economic life of the section than in other more industrialized areas. Add to this the popular view that the South is at once the most crude and the most courtly, the most promising, the most provincial, and the most backward of the regions of the United States.

The scientific interest in the culture of the South has resulted in many interesting and valuable interpretations of the region. To a social geographer like Ellsworth Huntington, the South is explicable in terms of climate; to publicists like Lothrop Stoddard and Madison Grant the explanation is wholly in terms of race; to an economic historian like Ulrich B. Phillips, in terms of race and the plantation system; and to a social historian like Walter L. Fleming, partly in terms of the results of Civil War and Reconstruction. Dr. David Starr Jordan stresses the biological depletion wrought by the Civil War. Professor Howard W. Odum has suggested a treatment of the contemporary southern situation in terms of leadership, and Dr. Edwin Mims has given an intellectualistic interpretation. Professors Holland Thompson and Broadus Mitchell have furnished historical commentaries on the South's industrialization of its cotton. Dr. Francis Butler Simkins and Dr. A. M. Arnett have traced attempts of southern rural folk in South Carolina and Georgia to better their conditions of life by political action. Frank Tannenbaum has published a brilliant though journalistic critique of the South in terms of cotton and racial attitudes. Possibly the most promising attempt at regional interpretation is the series of studies planned and under way in the Institute for Research in

Social Science at the University of North Carolina.[16]

Without assuming the position that the culture of the plant has done more than condition the development of the region, it is the purpose of the following chapters to trace the relation of cotton to its growers and the section. The presence of cotton is assumed in every study of social conditions in the South, and its culture is usually fitted into the predetermined categories of the investigation. The buyers and spinners speak of cotton in terms of supply. To them the question of cotton is a matter of increased production and decreased cost. To the county agent and the specialist of the experiment stations cotton offers a technical problem of increased production of better grades. To the experts of the Crop Reporting Bureau cotton is a matter of tabulation of acreages, plantings, weather damages, weevil damages, and ginnings, conditions of crops that are and that are to be. To the southern educator it is a matter of education. Given a school term of six months, eight months, nine months, all other things shall be added unto the children of cotton growers. To the racial propagandist cotton is simply another aspect of race, the exploitation of black men by white. To the inarticulate cotton farmer, cotton is often simply "hard luck." One year drought, another year weevil, a next year good crops and a market with the bottom knocked out of prices. Our concern is with the producers.

R. Clyde White in attempting to trace the relation of cotton to certain factors in southern culture used the method of correlation.[17] In eleven cotton states the 151

---

16 See Bibliography for writings to which reference is made.

17 "Cotton and Some Aspects of Southern Civilization," *Social Forces* (Sept., 1924), pp. 651–54.

counties having the highest acreage of cotton were taken. These counties represented 40.4 per cent of all the cotton acreage in those states for the census year 1919. The correlation of acreage planted to cotton with the extent of tenancy was found to be +.51, that of cotton and illiteracy was +.099, cotton and Negroes +.123, and tenancy and illiteracy +.537. In an area of high tenancy, illiteracy, and Negro population ratios the only significant relations found are those of cotton and tenancy and of tenancy and illiteracy. Mr. White concludes that the South cannot be regarded as a cultural unit nor can cotton, on the basis of mere correlation, be considered a causal factor.

In seeking to trace the reaction of cotton on its human factors in the South, one is confronted with the necessity of a unity of the social sciences. Descriptive and factual materials of human geography, economics, history, agricultural economics, and sociology accordingly are used wherever found valuable. In addition to the statistics the method of presentation by case studies has also been found valuable. Expressions of attitudes as found in interviews and letters, which would be regarded by historians as primary sources after a sufficient lapse of time, are also used.

From the methods and materials of geographers the facts of soil, climate, and topography may be employed to account in some measure for the concentration of cotton culture and the distribution of races, tenancy, and the plantation in the different areas of the Cotton Belt.

CHAPTER II

THE COTTON BELT: ITS REGIONS AND ITS HUMAN ECOLOGY

SPECIALIZED REGION

IN LESS than 3 per cent of the world's land area 60 per cent of the world's cotton supply is grown.[1] The Cotton Belt of the American South is thus one of the most highly specialized agricultural regions in the world. It contains about 295,000,000 acres, nearly one-sixth of the land area of the continental United States, extends 1600 miles in length, and averages 300 miles in breadth. In this area are 2,100,000 farms, one-third of the number in the United States,[2] which produced in 1919 crops valued at $3,800,000,000, one-fourth of the total farm income of the United States.[3] In this belt 42 per cent of the crop land was in cotton in 1919, and the value of the cotton crop was equal to the value of all other crops in the Belt combined.[4] As a matter of course cotton occupies the best land in the Belt, and the time devoted to other crops is determined by the demands of the cash crop. The value

---

1 Oliver E. Baker, "Agricultural Regions of North America," Part II, "The South," *Economic Geography* (Jan., 1927), p. 65.

2 These are, of course, census figures, but the census counts as a separate farm each tenant's holding in the South. This gives the South twice as many farms per unit of area as the remainder of the states. If each plantation were counted as one holding the South would fall behind in number of farms.

3 Baker, *op. cit.*, p. 63. 4 *Ibid.*, pp. 77, 79.

of cotton lint is exceeded by the value of corn, hay, and wheat crops for the whole United States, but when the value of cotton seed is included cotton ranks second only to corn.[5] In our export trade the value of cotton far exceeds that of any other commodity.[6] "It is the chief and often almost the only source of income to a large proportion of the farmers in the Southern States."[7]

The Cotton Belt is not one but many. Within this area differences in climate, rainfall, altitude, character of the soil, and history have given rise to subregions of cotton culture. These regions differ rather widely in the spatial distribution of what may be called human factors—black men, white men, share croppers, share tenants, small owners, and planters. Any adequate analysis of these regions would include the distribution in terms of regions, of population, races, types of tenure, domestic animals, cities, buildings, and machines devoted to cotton culture. This description of spatial distribution of man and the artifacts of his civilization in relation to cotton lands is, I take it, the human ecology of the Cotton Belt.

THE PLANT

"Cotton is the lint or fine fiber which grows on the seeds of plants belonging to the genus of *gossypium*. Wild species of *gossypium* are found in tropical regions of both hemispheres, and there are hundreds of cultivated varieties, differing in plant characters as well as in the length, strength, and fineness of fiber. Thirty-eight prin-

---

[5] A. M. Agelasto, et. al., "The Cotton Situation," Separate No. 879; also in Dept. of Agriculture *Yearbook*, 1921, p. 323.

[6] *Ibid.*, p. 324. See also J. Russell Smith, *North America*, p. 237.

[7] "The Cotton Situation" p. 324.

cipal commercial types are recognized at Liverpool."[8]

In the United States are grown three principal types: 1. Sea Island, staple 1⅝ to 2 inches in length, produced on the average about 100,000 bales annually, principally on the coast of South Carolina, Florida, and Georgia. This crop has been ruined by the boll weevil and in 1920 amounted to less than two thousand bales.[9] In 1926 Shepperson's *Cotton Facts* noted: "The only mention of Sea Island Cotton is to the effect that its cultivation has practically disappeared," since the area fitted for it has been completely overrun by weevil. 2. Upland long staple with a length of 1⅛ to 1¾, ranges about 1,500,000 bales per year, and sells up to 60 per cent higher than middling. 3. Upland short stapling is the standard "American Middling," ⅝ to 1 inch, which furnishes 92 per cent of the cotton crop.[10]

CLIMATOLOGY OF COTTON PRODUCTION

Although a native of the tropics, the cotton plant has become best adapted to the mildly tropic South. In the tropics it was a perennial; in the temperate climates it has become an annual. The limiting factors are climate, rainfall, and topography, with soil as a factor of lesser importance.[11] Cotton demands first of all a two-hundred day growing season, free of frost. This line is rather definite and follows the average summer temperature of 77 degrees closely.[12] The Cotton Belt has a rainfall almost twice as great as that of Illinois or New York. The

[8] "Cotton," *Atlas of American Agriculture*, Part V; Sec. A, p. 5. (This will hereafter be referred to as *Cotton Atlas*.)

[9] "The Cotton Situation," p. 329.

[10] *Cotton Atlas*, p. 3.

[11] Baker, *op. cit.*, p. 65.

[12] *Cotton Atlas*, p. 9.

limits thus set outline the traditional Dixie with remarkable fidelity. The line begins at the northern border of Virginia, slopes to the southern border of Tennessee, rises to the northern border of Arkansas and Oklahoma, and turns sharply to the south when it meets the arid regions of west Texas. The limiting factor in the area thus outlined is the amount of autumn rainfall. While it is true that heavy spring rains cause shallow rooting of the cotton plant and excessive summer rains restrict the yield, an autumn rainfall exceeding ten inches is practically prohibitive of cotton culture. This is because "frequent rain at this season of the year not only interferes with picking and damages the lint, but also favors destruction of the bolls by the weevil." [13] This has led geographers like O. E. Baker and J. Russell Smith to set off Florida and the subtropic gulf coast as a region separate from the Cotton Belt.[14]

THE COTTON BELTS: SOIL REGIONS

Within the limits set by frost line and rainfall the density of cotton production is determined by the soils and altitude. A dot map of cotton acreage [15] discloses four generalized areas of cotton production: the Eastern Coastal Plain, the Gulf Coastal Plain, the Central Alluvial Valleys, and the Western Prairie Lands. These areas are divided into subregions which correspond with definite soil areas.[16]

---

13 Baker, *op. cit.*, p. 67.

14 Smith, *North America*, chap. XIV.

15 See opposite page. Reprinted from *Cotton Atlas*, Plate 14, p. 9.

16 This analysis of the cotton belts follows the contribution of Hugh H. Bennett in his map prepared in 1917 for the Bureau of Soils, published in *Cotton Atlas*, Fig. 12, p. 8, and reproduced as

THE EASTERN COTTON BELT

The Eastern Cotton Belt is divided into five subregions differing in soil, characteristic, vegetation, and extent of cotton culture. In this region the cotton acreage begins at the southern border of Virginia and swings southwestward through North Carolina, South Carolina, and Georgia in two unequal strips, separated by the Sand Hills. This region is the oldest of the Cotton Belt, cotton culture in parts of the Carolinas dating back to colonial times. Many rivers fed by rains from the coast

---

frontispiece in the present volume. Many of the names given the regions are applied locally, and the changes from one to another can be noted by the casual traveler by automobile or train.

The figures of average yield per cotton acre were compiled for the *Cotton Atlas* from the census years 1879, 1889, 1899, 1909, while the figures of production of the acre are averages of the five years 1911–15. These averages are taken as being more nearly normal than any to be secured after the advent of the boll weevil.

The statistics of size of farm and type of tenure are taken from the atlas analysis of cotton production, studies of counties in the different regions. The number of typical counties selected for study in each is regarded as being enough to give an adequate sampling. Each of the regions has the following number of counties studied:

Regions	Counties Studied
Atlantic Coast Flatwoods	7
Upper Coastal Plain	16
Sand Hills	9
Piedmont Plateau	20
Black Prairies of Alabama and Miss.	7
Yazoo Mississippi Delta	10
Black Prairie of Texas	10
Interior Coastal Plain	9
	88

When the plantation system is mentioned the definition is that of the census—one continuous proprietorship divided into holdings among five or more tenants.

clay does not do so well. Potash is needed to hold the bolls on the plant in the heat of midsummer. High winds are likely to blow sand into the open cotton, producing the "sandies" much disliked by spinners.[20] The holdings are small, about 9 per cent of the land is in cotton and 24 per cent in plantations.[21]

The Piedmont Plateau is a clay belt whose characteristic red tinge can be traced, as Hubbard suggests, by a railway traveler all the way from New Jersey to the red clay hills of Georgia.[22] The vegetation is short leaf pine, oak, and hickory. The average yield of cotton is 180 pounds, and the area produces about 1,800,000 bales a year.[23] The hills and rolling surface prevent the long straight furrows of the coastal plain, making necessary curving rows that help in terracing. The cotton is the upland short staple, although better varieties are being introduced by experimental farms such as that of the Coker Company at Hartsville, South Carolina. The lower red lands need much less potash than the sand hills and coast land. In the upper Piedmont the cotton plants grow so short as to be called "Bumblebee" from the rustic quip that a bumblebee can stand on his hind legs and drink from the bloom. The farms average a little over thirty-five acres, about 20 per cent of the land area is in cotton, and almost 70 per cent of the farms are operated by tenants. The arrival of the weevil has pushed cotton culture north, and North Carolina has had the largest cotton yield per acre for several years. Her best producing cotton county lies in the Piedmont. In the southern tip of the red clay belt, southwest Georgia often produces the earliest cotton outside southwest Texas.

---

[20] Hubbard, *op. cit.*, pp. 4, 5. [21] *Cotton Atlas*, p. 12.

[22] Hubbard, *op. cit.*, pp. 4, 5. [23] *Cotton Atlas*, p. 12.

Much of this Eastern Belt has been planted to cotton for over a hundred years. This fact is responsible both for the lack of soil fertility and for the prevalence of the plantation which has been accepted as a heritage from slavery. A study of the expenditures for fertilizer in the United States shows the extent to which soil exhaustion has gone in this division. "The Eastern Cotton Belt, notably the Middle and Upper Coastal Plains and the Piedmont subregions, use more fertilizer than any other portions of the United States."[24] According to the 1920 Census, North Carolina, South Carolina, and Georgia spent approximately one and a half billion dollars on commercial fertilizer, almost as much as all the other states combined. Much of this expenditure is, of course, to be charged against tobacco culture.

THE GULF STATES BELT

From South Carolina in the Eastern Belt the Upper Coastal Plain swings through southern Alabama into central Mississippi. These two states comprise the Gulf section, the second oldest division of the Cotton Belt. It contains several varieties of soils that give rise to subregions. The Upper Coastal Plains reach almost to the coast where the sandy soil and excessive autumn rains prevent cotton culture. The rows often run straight across the field since the land is usually flat. The area consists of about twenty-eight million acres with rolling contour and soils of grayish to reddish sandy loam. The characteristic vegetation is pine, oak, and hickory. The farms average over forty acres, over 13 per cent of all the land is cultivated in cotton, and 28 per cent of the

[24] Baker, *op. cit.*, p. 71.

farms are run by the plantation system. The average cotton acre has produced 190 pounds, and the whole area produces on the average over a million bales.[25]

Just above is the bow-shaped region of clay hills, a continuation of the red hills of western Georgia through Alabama and Mississippi. The area is rather small, has hilly clay land with some "white rock" land, and produces about 320,000 bales on 8,000 acres of upland cotton.

A well-known subregion is the "Alabama Black Belt," so called from both its Negro workers and the crescent-shaped Black Prairies which curve upward from southeastern Alabama into northern Mississippi. The "black lands," moisty brown silt loams with post oak vegetation, grow a strong staple, 1 inch to 1⅛ inches in length. The humid climate and heavy soil have produced a type of cotton much sought after. Over one-half of the improved land area is in cotton, and 78 per cent of the farms are operated by Negro tenants, 64 per cent of the cultivated land being in plantations. The average holding is above thirty-five acres. The average "yield per acre is less than 150 pounds owing to continuous cropping and shallow plowing" mainly by unskilled Negro labor.[26]

A fairly productive small cotton area is the fertile Tennessee River Valley regions of northern Alabama with its brown and red hill soils. The Mississippi Bluffs extend from Louisiana into Kentucky. The Silt Loam Uplands are level and undulating and have suffered erosion, with the vegetation principally oak, sweet gum, and poplar. The cotton acre produces about two hundred pounds and the area produces about a half-million bales. The Gulf

---

[25] *Cotton Atlas,* pp. 8, 12.

[26] *Ibid.,* p. 8.

states have been hard hit by the weevil, and cotton acreage has not increased since 1912. The problem of weevil control is complicated by the wooded tracts which extend along the cotton fields and furnish places of hibernation for the insects.[27]

THE CENTRAL "RIVER BOTTOMS"

"Alluvial soils occupy a larger area in the Cotton Belt and subtropic Crop Belt than in any other region in the United States,"[28] and they are the most highly specialized in cotton. America drains itself into the Cotton Belt, sloping from the Appalachian table-land to the Atlantic in the East and through the vast interconnected Mississippi River system to the Gulf. An analysis of lands in need of drainage shows Florida, Louisiana, Texas, Georgia, Minnesota, Michigan, North Carolina, Mississippi, Arkansas, and South Carolina leading in acreage of wet, swamp, and overflowed lands. J. Russell Smith calls a chart of these areas "a map of the mosquito industry."[29] The course of the Mississippi, the Arkansas, and the Red rivers can be traced by the cotton production in these bottoms.[30]

The shift to the Central Valley had begun by 1821, and in 1833 less than half of the cotton production was in the Atlantic Coast. The first areas developed in the Alluvial Bottoms of the Mississippi were around Memphis and in the delta formed by the Yazoo where it flows into the Mississippi. The brown and mottled clay soil characterized by cypress, red gum, and oak growth produces a smooth silky staple as long as 1¾ inches and of ex-

[27] Hubbard, *op. cit.*, p. 8.
[28] Baker, *op. cit.*, p 73.
[29] *North America*, p. 168.
[30] *Cotton Atlas*, Fig. 14, p. 9.

ceptional strength.[31] The occasional overflows to which the land on the three rivers mentioned is subject serve to enrich the soil. Old buyers have given special names to the best cotton grades. "Benders" are grown in the bends of the Mississippi; "rivers," on the banks of tributaries to the Father of Waters, and "creeks," along the smaller streams.[32] "The size of the yield and the height and vigor of the plants are exceptional." Hubbard speaks of old-time photographs displayed in many cotton offices showing a planter on horseback in his field with the animal almost hidden in the foliage.[33] The Yazoo Delta has the highest average yield of cotton in the Cotton Belt, 265 pounds to the acre. In the Delta 70 per cent of the improved land is in cotton, 85 per cent of the farm land is operated according to the plantation system, and 86 per cent of the farms are operated by Negro tenants. The average holding is about thirty-three acres with around twenty acres in cotton. The sixteen million acres average about a million bales.[34]

The so-called second bottoms, which lie above overflow, produce good yields in southeastern Missouri and northeastern Arkansas. Very little fertilizer is found necessary on any of the alluvial farms.[35] The weevil has also spread devastation in the alluvial valleys, playing especial havoc with fine delta cotton.

The plantations in Arkansas and Mississippi are much larger than those found elsewhere. The largest plantation in the world at Scott, Mississippi, in the Delta contains 37,000 acres. The land is flat, and the rows stretch far away. One viewing the region for the first time is

---

[31] Hubbard, *op. cit.*, p. 8. [32] *Ibid*, pp. 10–11.
[33] *Loc. cit.* [34] *Cotton Atlas*, pp. 8–12.
[35] "The Cotton Situation," p. 348.

likely to be oppressed by the lowness of the country and the innumerable Negro tenant shacks, each with its cotton house, that stretch away into the distance.

THE WESTERN BELT

The movement to the Western Belt came comparatively late. Texas was added to the list of cotton states in 1845 and in the statistics of 1859 and 1869 ranked fifth in production. By the census of 1890 the state had assumed the leading place in the Cotton Kingdom and cultivation was well under way in the Indian Territory. Texas and Oklahoma are regarded as the frontier of the Cotton Belt, having been but recently reclaimed from the long and short grasses of prairie and plains. Together they comprise one-sixth of the area of the Cotton Belt.

"The general characteristics of Texas and Oklahoma are a rich and alluvial soil belt with stretches of poorer land on the outskirts, but everywhere a rolling country like the western prairies."[36] The cotton produced in this area is likely to have a staple longer and stronger than that of the East but not equal to that of the valley.[37]

The leading subregions are the Interior Coastal Plain, Eastern Oklahoma and Red Prairies, the Black Waxy, the Grand Prairie of Texas, and the Great Plains.

The Interior Coastal Plain extends through Northwest Louisiana, Southwest Arkansas and Northeast Texas. The topography is rolling, the soil, grayish and reddish sandy loam, the vegetation, pines and oak giving way to prairie area in Texas. One-third of the improved land is in cotton, and almost half of the farms are operated by owners. An acre of cotton produces about 165 pounds

[36] Hubbard, *op. cit.*, p. 12. [37] *Ibid.*, p. 13.

and the whole area averages almost a million bales.

The Eastern Prairies of Oklahoma comprise almost half the eastern part of the state. The soil is black to reddish, topography rolling, and vegetation prairie grasses. The area is mostly given to corn, hay, and pasture land. In 1909 it produced over a third of a million bales averaging 182 pounds to the cotton acre.

The Red Prairies in Western Oklahoma and North Central Texas are rough in contour and given mostly to grazing, produce about 105 pounds of cotton to the acre, but bid fair to become important because of large area. They produced 825,000 bales in the 1909 figures.

The outstanding cotton section of the West is the Black Waxy of Texas, so called from its dark calcareous clays. It has the highest per cent of its area in farms, 86, the highest per cent improved, 62, and the highest in cotton, 31.6. It is also unique in that it is an area of white tenancy, for although only 14.5 per cent of the farm land is in plantations, 55.7 per cent of all farms are run by white tenants. In the per cent of white ownership the region comes second only to the Interior Coastal Plains. The size of farms is over sixty-five acres, second to the Red Prairies with their hundred-acre farms—much of which is in pasture. The land is rich and fresh and practically no fertilizer is needed. West Texas has a potentially fertile soil but the desert is too close and the "droughts are frequently devastating."[38] The story of the spread of cotton culture to the Great Plains must be left to a later chapter. The black land, however, does not suffer so much from drought, for "it holds water like a sponge."[39]

---

[38] Hubbard, *op. cit.*, p. 79.

[39] *Ibid.*, p. 15.

"COTTON AND 'CAWN'"

A spot map of the distribution of corn shows that it is cultivated in all portions of the Cotton Belt although there is no such concentration as found in the Corn Belt reaching through Iowa, Illinois, and Indiana. Baker estimates that 48 to 52 per cent of the crop land in the Cotton Belt is given over to the production of feed for live stock.[40] This includes cotton seed, however, which is estimated to occupy about 6 per cent of the crop area. In areas of less intensive specialization corn often occupies an area equal to that of cotton. The farming system is based upon the two crops which are "planted in succession or alternation until the productiveness of the soil is reduced."[41] The Belt does not produce enough corn to supply its own needs. As a result the price of corn in the Cotton Belt is often twice as high as in the Corn Belt. On the other hand, the production per acre is about half that of the Corn Belt. Dr. Baker suggests that "the production of corn, hay, cattle, and hogs in the Cotton Belt appears likely to be profitable only up to that point at which the supply of these products does not exceed the local demand."[42] H. C. Taylor also suggests that corn may pay when grown for home use and yet fail to achieve a place as a commercial crop.[43] The northern and western fringes of this Belt are devoted to combination crops of wheat, rye, and oats. These are exceeded in acreage further south by cow peas, velvet beans, and peanuts. Sweet potatoes are evenly distributed throughout the humid portions of the whole Cotton Belt. The facility and abundance with which these "by-products" crops can be

---

[40] Baker, *op. cit.*, p. 79. [41] *Ibid.*, p. 80. [42] *Ibid.*, p. 82.

[43] *Outlines of Agricultural Economics*, p. 60.

grown in the South is responsible for the fact which W. E. Dodd notes when he says that "the problem of subsistence during the Civil War was much simpler than in any of the European countries fighting in the Great War." [44]

Rainfall, however, limits southern agriculture in the production of hay. The higher rainfall for all except the Western area of the Cotton Belt results in autumn showers that make the growing of hay extremely hazardous. "In all the states from North Carolina to Louisiana many farmers attempt to produce hay," but a large part of the crop is lost each year in the curing and much of the rest is damaged by untimely rains. More dependable autumn weather is found in states further from the Atlantic and the Gulf, particularly Arkansas, Oklahoma, and parts of Texas, all of which have larger acreages of hay. As a result, the relative proportion of crop land devoted to hay in the cotton states is the smallest in the United States. Not enough hay is produced for home use, and much has to be imported from the Middle West. Since the freight charges often equal the original cost, southern planters pay a high price for their hay. If methods of curing suitable to the climate were worked out, hay would be more generally grown. It would, however, be valuable as a cash crop if only a few farmers in each community grew it for market.[45]

Excessive rainfall is the main regional factor responsible for ruining a great deal of southern farm land. It is estimated that in the Cotton Belt in the decades since 1860 "erosion has destroyed an area equal to that

[44] *The Cotton Kingdom,* p. 63.

[45] W. J. Spillman, *Distribution of Types of Farming in the United States,* Farmers' Bulletin 1289, p. 12.

of Belgium." Regional factors combine with the tenancy system to allow the fields to wash away. Southern soils are peculiarly liable to leaching. Iowa, for instance, has thirty inches of rainfall a year, and her soils are frozen impervious to water all winter; the cotton states have fifty to sixty inches of rainfall and winters that are nearly frostless. Grass serves to retain the soil, but cotton and corn are two crops which require that vegetation be weeded out. Thus agriculture without grass and without humus has been combined with a tenant-landlord system to ruin thousands of acres. Gullies best described in the term "red washes" have reduced many of the best upland farms to sandy wastes.[46]

MULES

Corn, mules, cotton is a logical sequence, and no ecology of the Cotton Belt is complete without a study of the distribution of mules. Spot maps prepared from the Census show that "there are more mules and fewer horses per square mile in the Cotton Belt than in any eastern agricultural region." [47] The ten ranking states in mules are all southern. The horses are localized in the Corn Belt and with the exception of Missouri and the cotton frontier of Texas and Oklahoma, comparatively few mule colts are found in the cotton states. Missouri, Kansas, Oklahoma, Texas, Tennessee, Kentucky, Iowa, and Nebraska rank highest in mule colts. More pasture, cheaper feed, better methods of stock-raising have arranged it so that mules are grown in the Corn Belt and shipped South when ready for work.

---

[46] See Smith, *North America*, p. 255. [47] Baker, *op. cit.*, p. 80.

CITIES BUILT ON COTTON

The location and growth of southern cities have been determined largely by their relation to cotton culture. The human ecology of the Cotton Belt would include an account of the agricultural capitals, the distribution of cities built on cotton. Cities in relation to cotton are divided into future markets, spot markets, and points of export. A spot market may be either bona fide or used for determining difference for delivery on future contracts. These do not include the primary markets, the name given to the interior towns and villages where the cotton first leaves the hands of the producer.[48] For the Eastern Belt, Savannah, Norfolk, Augusta, Charleston, Atlanta, Columbia, Brunswick, Wilmington, Macon, Greenville, lead as spot markets in the order named. Savannah, Brunswick, Wilmington, and Norfolk are also export points. For the Gulf states, New Orleans, Mobile, Montgomery, and Pensacola, Florida, serve as outlets. The alluvial regions form the hinterland for the great cotton metropolises of New Orleans and Memphis, with Little Rock, Pine Bluff, and Helena as smaller Arkansas spot markets. New Orleans is remarkable as the cotton outlet of the Mississippi. Third as a spot market in receipt of cotton bales, second after New York as a future market, and second as a port of export, New Orleans is a really great cotton metropolis.

The great Western Belt is dominated by Galveston and Houston, which lead the world as spot markets. Texas City, Fort Worth, Paris, Dallas, and Chickasha, Oklahoma, are also cotton centers. Galveston leads as point of export. The leading cotton points outside the

[48] "The Cotton Situation," pp. 383-85.

Belt are St. Louis, Norfolk, and Baltimore on the fringe, Boston as a center for the textile industry, Washington, San Francisco, and Philadelphia as points of export, and New York and Chicago as future markets. The remarkable expansion of the acreage in the Western Belt is shown by the growth of Dallas. From 1915 to 1920 Dallas received on the average 117,179 bales of cotton. In 1923–24 with the aid of eight steam railways and six electric lines Dallas handled a half-million bales. Three hundred million dollars were required to finance the transactions.[49]

The distribution of cotton warehouses in 1921 shows a distinct concentration in the Eastern and Western Belts, especially northern Georgia, South Carolina, and eastern Alabama. The Black Prairie of Texas also is well supplied with places of storage. The warehouses are found "at many local markets as well as at the larger concentration points throughout the South." [50] The noticeable lack of warehouse facilities for cotton is found in the great Mississippi River Bottoms where cotton is customarily marketed, due to the tenant system, as soon as gathered. In 1921 there were 2,735 warehouses in the South, Georgia with 775, Texas with 756, Alabama with 282, South Carolina with 269. The distribution of cotton compresses affords an interesting contrast. In the West, where most of the cotton is exported or shipped to New England, compresses are numerous; while in the East, where much cotton is consumed in the local mills, they are few.[51] Cotton compresses are of necessity located in the great urban centers, the spot markets and points of export. In the Western Belt, Houston, Galveston, Waco,

[49] "Dallas as Cotton Market," *Commerce and Finance* (Sept. 17, 1924), pp. 1801–5.

[50] "The Cotton Situation," p. 377. [51] *Cotton Atlas*, p. 25.

Dallas, and Fort Worth; in the Valley, New Orleans, Memphis, Little Rock, and Greenwood; in the East, Wilmington and Charleston are points at which bales are compressed for long distance shipment.

The cotton gin is the typical stage property of the southern small town scene. The whole Cotton Belt is covered with active ginneries. The area of greatest concentration of cotton gins is a comparatively uninterrupted stretch from North Carolina through the Alabama Black Belt. The Mississippi fringe comes next. In the western area gins are a good deal less numerous in comparison with the volume of production.[52] This is owing to the fact that the gins are of a later type, larger, and of greater capacity. In 1914 gins in Texas averaged slightly over 1,000 bales each, while those in South Carolina averaged only 490 bales. In 1915 the total number of active gins was 24,547, which decreased to 15,459 in 1925. The average number of bales ginned per active establishment, on the other hand, increased from 526 to 1,042. Texas, as would be expected from its size, leads with 3,459 gins. South Carolina, Georgia, North Carolina, Mississippi, Oklahoma, and Arkansas lead in the order named.

If the gin dates back to Eli Whitney, the cotton oil mill is a much more recent factor. Cotton oil mills have been erected only since the utilization of the seed, beginning in 1833. In 1915 only about 84 per cent of the cotton seed produced was crushed. The center of the industry is Memphis. The distribution of the cotton oil mills follows the outline of the Belt generally with the tendency to localization in larger cities.

The distribution of rural and urban population re-

---

[52] *Ibid.*, p. 24.

mains to be considered. The place map of urban population shows, of course, a concentration in the upper northeastern segment of the United States. But the similar map for country population shows almost as great a density in the South, east of the Mississippi River, as in the East and North. The densest country population in the South clearly follows the outline of the mountain section of eastern Tennessee, Kentucky, and western North Carolina. "There, although only a small part of the land is cultivated, the population is denser than in Illinois and Iowa where practically all the land is farmed."[53] Country population, however, is much more concentrated in the Cotton Belt than in the Corn Belt. In the West the Texas Black Lands stand out because of dense rural population. The mapping of villages shows concentration in the East, and near the areas of large cities. There is very little discernable relation to the configurations of the various cotton belts except that the Black Prairie again stands out in Texas. The Alabama Black Land Belt and the Mississippi Bottoms show a sparsity rather than a density of villages.[54] The distribution maps of Negro and white rural population[55] indicate that the Negro follows the contour of the areas of cotton production to a remarkable extent. The Eastern Belt, the Alabama Black Lands, and the Mississippi River Bottoms again stand out. The Negro population extends outside the Belt northward along the eastern fringe, farther into the rural districts of Virginia, Maryland, and Kentucky. The Texas and Oklahoma cotton areas are notably given over to native whites. The density of the white popula-

---

53 "Rural Population and Organization," *Atlas of American Agriculture,* Part IX, Sec. I, p. 4.

54 *Ibid.,* Fig. 4, p. 5. 55 *Ibid.,* Figs. 10, 13, pp. 7, 8.

tion is noticeably sparser in the warm and humid areas of the Cotton Belt, but appears concentrated in the mountainous areas of the Appalachians, Ouachita, and Ozark Mountains where cotton is not cultivated.

In 1910 the Census of the United States made a survey of plantations in the South for selected areas using a special schedule. For purposes of the survey the Census Bureau adopted the following definition of the "tenant plantation":

> A tenant plantation is a continuous tract of land of considerable area under the general control or supervision of a single individual or firm, all or a part of such tract being divided into at least five smaller tracts which are leased to tenants.[56]

Plantations were found to occupy an important part of the farming area of nine cotton states. A dot map of the plantations [57] in the counties selected for the survey shows in clear outline the Eastern Belt, the Mississippi-Alabama Black Prairie, the Mississippi River Bottoms and the Black Waxy of Texas. More plantations are found in the eastern areas where slavery was well developed before the War and where the soil is fitted for intensive cotton culture.[58] In the migration to the Alabama and Mississippi regions planters carried their slaves, but the slavery régime was not fully transplanted to Texas before the outbreak of the Civil War. Consequently, although the Black Waxy has a high percentage of tenancy, it differs from all the other areas in being overwhelmingly white tenancy. White tenants operate 55.7 per cent and Negro 9.5 per cent of farm lands in the

[56] "Plantations in the South," 1910 *Census*, V, 878.

[57] *Cotton Atlas*, Fig. 32, p. 11. [58] Baker, *op. cit.*, p. 84.

Texas Black Lands. With this exception the dot map of cotton farms operated by Negro tenants follows the plantation map accurately. The map of distribution of Negro owners shows them sparsely scattered all over the South. "They are relatively most numerous on the Atlantic Coastal Plain and the Interior Coastal Plain in Arkansas, Louisiana, and Texas. They are least numerous where the plantation system prevails."[59]

While cotton plantations are found in all the cotton states, sugar cane, rice, and tobacco plantations are found confined to special limited areas. Tobacco and cotton are cultivated together on plantations in the Coastal Plains of North and South Carolina. Rice planters with a developed tenant system occur in east central Arkansas, Louisiana, and southeast Texas. The sugar cane plantations are found in southern Louisiana.[60]

A mapping of the white tenants[61] shows them most numerous outside the regions of the plantation régime. They are concentrated mainly in the upland regions such as Piedmont, northern Alabama and Mississippi, the Ozark and Ouachita Highlands except in the prairies of Oklahoma and Texas, where they are fairly numerous. The white owners are thickly scattered over all the South, being possibly more concentrated in the upper Piedmont due to the small size of farms.

It is not to be rashly disputed that the cotton plant is something of a map maker. As cotton has thrived in various soils and areas it has brought about the distribution of races through the movement of slaves with the

[59] *Cotton Atlas,* p. 12.

[60] C. O. Brannen, *Relation of Land Tenure to Plantation Organization,* Dept. of Agriculture Bulletin 1269, p. 8.

[61] *Cotton Atlas,* p. 13.

masters. Systems of land tenure have followed the migration of cotton culture to the Western Belt. Gins, compresses, oil mills, and warehouses have been built to facilitate the handling of the fiber and its seed. As cotton culture has increased cities have grown up to serve as primary markets and points of export. Over and above all is the warm sun that makes cotton the product of the South. Under all is the soil which seems by its texture and topography to sift out races and systems of land tenure to fit with cotton culture.

CHAPTER III

THE EVOLUTION OF THE COTTON SYSTEM

THE NATURE AND SPREAD OF THE PLANTATION SYSTEM

THE COTTON system may be defined as the complex organization of financing, growing, and marketing cotton. It includes croppers, tenants, small farmers, and planters who comprise the growers, plus the banks, supply merchants, factors, and fertilizer dealers who finance the crop and the coöperatives, local buyers, general buyers, shippers, and exporters who assemble and classify cotton for sale to the mills or for export. The cotton system as it exists today is complex and far-reaching. Men who have dealt in cotton for years frankly admit that phases of the business exist which they have not fully explored. The history of the cotton system is divided distinctly into two parts by the abolition of slavery. The present cotton system is an adjustment, man to land and race to race, that has been received as a social heritage from the past. No one can presume to understand the practices in the production of cotton or its tremendous hold on the South without viewing them as an evolution from the days of the Cotton Kingdom and slavery. Until the Civil War the history of cotton production was practically one with the history of the plantation.

One characteristic feature of the material culture of the South throughout its long course of development has been the plantation system. The economic and social ad-

As Phillips has said:

> This usage of the word in the sense of a colony ended only upon the rise of a new institution to which the original name was applied. The colonies at large came then to be known as provinces or dominions, while the sub-colonies, the privately owned village estates which prevailed in the South were alone called plantations.[5]

The system of land grants made the plantation inevitable. Holdings of land were offered to colonizers in proportion to the number of settlers they could bring over. This was the method of arranging for the transportation of settlers. It also fitted in with a feudal society's conception of landed estates. One proprietor, Lord Baltimore, offered a thousand acres for every five settlers brought over. It was understood that the land was to be rented to the men who were brought over, and the rent was to consist of a part of the produce, usually tobacco. Thus we see that as early as 1636[6] a system of share renting on large holdings had been introduced, and only awaited the advent of slaves to become the plantation system.

The industrial history of America, like that of all new countries, is to be understood in terms of rich natural resources, scanty capital, and a labor supply totally inadequate for exploitation. It can easily be seen that the developing of tobacco, rice, indigo, sugar, and cotton culture would impress more and more upon the South the demand for labor. The Virginia Colony around Chesapeake Bay had dragged out a most precarious existence

[5] *American Negro Slavery*, p. 309.

[6] W. B. Bizzell, *Farm Tenantry in the United States*, pp. 162–63.

until John Rolfe introduced the cultivation of tobacco in 1612. "About the last of August," he wrote in 1619, "came in a dutch man of warre that sold us twenty negroes." Tobacco soon came to be the staple export crop of the new world. Natives of Guinea, "very loyal and obedient servants, without malice," who "never more tried to fly, but rather in time forgot about their own country" had been inducted into the plantation system. These plantations were located on all the fertile river banks so that the tiny ships of colonial days could sail up and take a cargo from each farmer's wharf.[7]

The small grain and fishing culture of New England, as has often been recounted, discouraged slave labor. Later the rise of the urban-industrial culture in the northeastern states, having accumulated capital goods, found its labor supply in "thousands of immigrants who came at their own expense, who worked zealously for wages payable from current earnings, and who possessed all the inventive and progressive potentialities of European peoples."[8] In the meantime, the South had capitalized her labor supply in slaves so that she owned both her laborer and his labor. This gave to the free artisan and agricultural laborer such a low social as well as economic status that the new immigrant failed to seek out the South and even the native farmer became restless and in many instances removed to the mountains, pine forests, or further to the west.

The system under which the labor force was apportioned to production of southern agricultural staples came to be known as the plantation system. The plantation is an application of the capitalistic system to agri-

---

[7] Faulkner, *op. cit.*, p. 67. [8] Phillips, *op. cit.*, p. 395.

cultural production and possesses the characteristics of American extensive farming. Where lands exceed labor supply large scale capitalistic agricultural production may be found in the frontier cattle ranch with its hired cow-hands. Later in the stage of development wheat farming emerges with its use of machinery and casual laborers. Slavery and the plantation system differ from serfdom and the manor system in that the slave labor supply is more mobile. Not being attached by law and custom to the land the slave was present with his master in the march of the Cotton Kingdom westward.

The elements of the plantation system are four: First in order comes a land supply of large acreage of fertile soil, cheap, level or rolling, and to some extent homogeneous in texture and topography. Prairie lands and river bottoms are suitable for the plantation; there are no records of its being applied to mountainous regions. For such land a labor supply of low social status, docile, and comparatively cheap is desired. In the third place the management required is social as well as economic supervision. And last, the products must be staples, routine crops easily cultivated by set rules; cash crops for which no problem of marketing exists.

The South has furnished five such crops, routine in production and staple in demand. "It would be hard to over-estimate the predominance of the special crops in the interest and industry of the southern community. For good or for ill they have shaped its development from the seventeenth to the twentieth century. Each characteristic area had its own staple, and those districts which had none were scorned by all typical southern men. The several areas expanded and contracted in response to fluctuations in the relative price of their products." [9] Tobacco, indigo, sugar, and cotton are all adapted to gang labor working on time basis under an overseer. Rice, however, has been cultivated on a task-work basis with so much assigned to each slave. All these staples fought a losing battle with cotton.

The scarcity and cost of white labor led by degrees to the introduction of slaves. Convicts and indentured servants became more and more things of the past. Land was easily acquired; great plantations "often had a thousand acres under actual tobacco cultivation," many were over five thousand acres in size. "Fifty acres of arable land per Negro were considered necessary for profitable cultivation, and an overseer was too expensive unless he had twenty Negroes under him." [10] Tobacco exhausted the land in three years and forced the use of other crops.

In the coastal Carolinas and the lower South the plantation system grew up about rice and indigo culture. Rice culture, introduced from Madagascar into the Carolina swamp lands, did not prove profitable until Negroes were introduced to work in the hot wet fields. Charleston became the export point of a great rice area. Indigo was introduced by Eliza Lucas, daughter of an English army officer, on her father's plantation. A bounty aided in its production, and in 1775 the value of indigo exported was over one million pounds.

In the decade beginning 1783 a widespread depression in the plantation system prevailed. The production of indigo was in its decadence, rice cultivation was changing to the new tide-flow system, prices were low, and each new tract opened to the plantation system meant an old one

[9] *Ibid.*, p. 205. [10] Faulkner, *op. cit.*, p. 67.

abandoned.[11] Only sugar had succeeded both with the plantation and the market.

Successful sugar culture first began in the Mississippi Delta below New Orleans when "Etiene de Bore, a prominent Creole whose estate lay just above the town, bought a supply of seed cane from Silis (who was making sugar with indifferent success as early as 1791), planted a large field with it, engaged a professional sugar maker"—and sold his 1796 crop for some $12,000.[12] The triangular district created by the confluence of the Red and Mississippi rivers became, in time, the scene of great sugar plantations stretching out along behind the levees. At the greatest height of the sugar industry in 1849 the plantations numbered 1,536, their slaves over a hundred thousand, and in 1853 they produced 450,000 hogsheads of sugar.[13]

A backward glance over southern industrial history serves to convince the student that the South awaited only the advent of cotton to extend the plantation system far and wide. Cotton found the beginnings of the plantation régime established but waning. Tobacco was proving too exhausting to unfertilized soils. Jefferson wrote in 1781 that the culture of tobacco "was fast declining at the commencement of this war" and "it must continue to decline on the return of peace." [14] The culture of indigo and rice were both on the decline. The South itself was lukewarm on the subject of slavery; all the states except one had abolished the slavery trade; and except for the sugar interest the plantation system was stagnant. It has been suggested by many historians

---

[11] Phillips, *op. cit.*, p. 150. [12] *Ibid.*, pp. 163–68. [13] *Loc. cit.*

[14] Cited by J. A. B. Scherer, in *Cotton as a World Power*, p. 147

that the abolition of slavery by these states would have been in course of time a natural and an easy process.

Cotton found the plantation system on the decline; it revived and pushed this system across the southern map. In 1793, the year Whitney invented the gin, the South produced 10,410 bales of cotton and exported 3,565; in 1810 production had risen to 177,824 bales and exports were 124,116. The Reverend William Winterbotham wrote in 1795: "Cotton has been lately adopted as an article of culture in the southern states; and as the prices of rice, tobacco, and indigo decline, it must be very beneficial."[15] "This economic transformation," says Frederick J. Turner, "resuscitated slavery from a moribund condition to a vigorous and aggressive life."[16]

The production of cotton awaited a method of separating the lint from the seed. In the Carolina tidewater the problem was already solved as early as 1791 to 1794 by growing the two-inch lint of sea island cotton which could easily be separated from its smooth, black seeds by rollers. Such cotton commanded fancy prices, increased the number of coastal plantations, and made their owners rich. But sea island cotton was limited to a definite area. By 1797 the cotton gin, invented in 1793 by young Mr. Eli Whitney, Yale graduate, was in operation in as many as thirty points in Georgia alone. The culture of short staple cotton soon spread over the Carolina-Georgia uplands. Lacking any staple crop, the uplands eagerly adopted the production of cotton and with it, to a fair extent, the plantation system. Phillips measures the extent of the planters' régime by the gain of the upland counties of South Carolina in Negro population from

[15] A. B. Hart, *American History Told by Contemporaries,* III, 67.
[16] *Rise of the New West,* p. 49.

less than one-fifth of the white population in 1790 to almost fifty per cent in 1830.

The farmers with a slave or two greatly outnumbered the planters and one "shrewd contemporary observer[17] found special reason to rejoice that the new staple required no large capital and involved no exposure to disease. Rice and indigo, he said, had offered the poorer whites except few employed as overseers, no livelihood without the degradation of working with slaves, but cotton stimulating and elevating these people into the rank of substantial farmers tended to fill the country with an independent, industrious yeomanry." Such is the irony of historian turned prophet.

The westward movement of cotton and the plantation followed as a matter of course. "The Alabama-Mississippi population rose from 40,000 in round numbers in 1810 . . . to 1,660,000 in 1860, while the proportion of slaves increased from forty to forty-seven per cent."[18] During the same period the Delta lands of Arkansas and Louisiana were filled with cotton planters. From 1815 to 1860 was the heyday of the plantation system. Indigo had seen its day, hemp was negligible, sugar culture was growing, tobacco while losing in the East was gaining in the West, but over and above all, in uplands and river bottom, from the Carolinas to Texas, Cotton was King! From 1791 to 1860 the average annual production had risen from in round numbers five million to 1,750,000,000 pounds; the exports from 1,740,000 to 1,390,000,000 pounds; and the per cent of the crop exported from 33.0 to 79.5. The plantation system was truly a Cotton Kingdom.

---

[17] David Ramsey, whose *History of South Carolina,* II, 448–49, is cited by Phillips in *American Negro Slavery,* p. 160.

[18] Phillips, *op. cit.,* p. 171.

It is difficult to overestimate the influence of the plantation régime. The plantation was the basic feature in the economic life, the planters were the capstone of the social system. It has been maintained by a leading historian of the South, that "nine-tenths of the South's landowners at any period in her history were small proprietors." [19] The same authority estimates that "three or four thousand families . . . lived on the best lands and received three-fourths of the returns from the yearly exports. Two-thirds of the white people of the South had no connection with slavery and received only a very small part of the returns of the community output. A thousand families received over $50,000,000 a year, while all the remaining 660,000 families received only about $60,000,000." [20]

We know, however, that the small farmers outside the plantation system lived in the valleys of Virginia and the uplands and highlands of middle and western North Carolina, north Georgia and Alabama, east Tennessee and Kentucky, and western Virginia. Under an economy more or less domestic and self-supporting they raised cereals, tobacco, cotton, and live stock. In the Eastern Belt their natural markets were Baltimore and Philadelphia. Their products were transported over the rough back-country highways, along which they brought their purchases. In the Western Belt the Mississippi River and its tributaries were used as highways to market. The planters, although outnumbered, admittedly retained control in politics. Socially the two classes had much in common; their ideals were often southern to provincialism. Eco-

---

[19] W. E. Dodd, "Plantation and Farm System in Southern Agriculture," *South in the Building of the Nation*, V, 74.

[20] *The Cotton Kingdom*, p. 24.

nomically the isolation of the two great classes was complete. Sectional contests developed; the small farmers were more than willing to vote taxes upon the great planter to build roads and canals in and out of the uplands. The interior often received population renewals from the planter section of unpropertied southerners who had given up the unequal struggle with slave labor and struck out westward.

Even these people were to find themselves crowded by the plantation system for "the rapid growth of the short staple cotton industry was responsible for the spread of the planter régime over most of the fertile hill country of South Carolina, Georgia, Alabama, Mississippi, Middle Tennessee, North Louisiana, and Lower Texas."[21] The new steam railways built before the Civil War gave cotton an outlet to the market and made its conquest of the regions permanent.

No less important than the spread of cotton culture is the structure and internal economy of the plantation. The technique of cotton production, the disposition of the human factors in production, their living standards, the management and relation of the plantation economy to the outside world are all of significance. They may best be treated by giving as far as possible cases and types that strike near the mode of plantation activity.

William J. Barbee of De Soto County, Mississippi, writing immediately after the Civil War has described the organization of a small plantation.

"The best bottom plantations," he writes, "are those immediately on rivers above overflow. Such location is decidedly healthier than any in interior of the bottoms." Such a plan-

[21] Dodd, in *South in the Building of the Nation,* V, 77.

tation must have good soil that will not wash away, good timber and plenty of it, good water in abundance, and must be close to a good steamboat landing or depot. Such bottom land may be counted on to grow one to two bales per acre to only one to a half bale for upland. Nevertheless, "for a family residence and plantation we think the best table and creek-bottom land of the hill country is more desirable—more especially when we take in view the social and moral advantages.

A good small plantation he regards as consisting of 200 acres. To stock such a farm will require the following supplies:

4 horses or mules at $150	$600.00
4 turning ploughs	25.00
4 broad shovels	25.00
plough harness	25.00
1 wagon	80.00
1 yoke oxen	100.00
axes, hoes, shovels, spades	20.00
saws, augers, chisels, hammers, tools	25.00
cross cut saw	10.00
	$910.00
Corn from Jan. to Sept.	300.00
	$1210.00

To run this plantation will require "six good hands in the field" and two in the house. Under favorable circumstances the plantation will produce about 35 bales of cotton and 800 bushels of corn. With cotton at a good price one can meet the expenses of the hands, keep up his family, have enough corn for the stock for next year, and possibly a large enough surplus to pay for the stock and implements.[22]

22 *The Cotton Question*, pp. 81-83.

Mr. Barbee's estimate may be taken as average for the type of soil and locality presented. Many of the plantations were much larger in size, but an estimate may be arrived at by multiplying the stock and number of hands, leaving the number of domestic servants more or less constant.

The régime on the plantation was well ordered. The slaves had their quarters, small log cabins or shacks of one thickness of board, built together near the big house. The movements were regulated by the plantation bell. The schedule on the Hammond plantation of South Carolina is cited as a type:

> The first horn was blown an hour before daylight as a summons for work hands to rise and do their cooking and other preparations for the day. Then at the summons of the plow driver at the first break of day, the plowmen went to the stables whose doors the overseers opened. At the second horn, just at good daylight, the hoe gang set off for the field. At half past eleven the plowmen carried their mules to a shelter house in the fields, and at noon the hoe hands laid off for dinner, to resume work at one o'clock, except that in hot weather the intermission was extended to a maximum of three and a half hours. The plowmen led the way home by a quarter of an hour in the evening, and the hoe hands followed at sunset.[23]

The food and clothing for the slaves were simple, substantial, and monotonous. De Bow estimated that slaves could be supported for $15 a year. Hoecakes, fried pork, and molasses formed the staples of diet.

> On the Telfair plantation in Georgia each worker was allowed a peck of corn, a pint of salt and not over three

[23] Phillips, *op. cit.*, p. 268.

and a half pounds of meat a week. Hammond also gave order to allow a heaping peck of meal and three pounds of bacon with the substitution in the winter if desired of a bushel of sweet potatoes for a peck of meal. Molasses was furnished in proportion. "Feed everything plentifully but waste nothing," was the admonition of one planter.

On the Hammond plantation each man was allowed two cotton shirts, a pair of woolen pants and a jacket in the fall; in the spring he drew two cotton shirts and two pairs of cotton pants. For the women there were six yards each of cotton and woolen cloth in the fall and twelve yards of cotton cloth in the spring with needed buttons, needles and thread. Each worker was to have a pair of stout shoes in the fall and a heavy blanket every third year. Negroes must appear once a week in clean clothes, "and every negro habitually uncleanly in person must be washed and scrubbed by order of the overseer—the driver and two other negroes officiating." [24]

The most difficult task in the cotton régime fell to the lot of the plantation overseer.[25] On him was the burden of the direction and integration of the plantation's activities. If he has been painted as an unlovely character it is largely because he was placed in a most unlovely position. General manager of the plantation, intermediary between master and slaves, he possessed the respect of neither. In the absence of the planter he was the only white man among a horde of blacks. But whether the planter was present or absent, the overseer was permitted intimacy with neither the planter nor his slaves. Up in the morning before anyone else to ring the planta-

[24] *Ibid.*, pp. 265–66.

[25] In *The Plantation Overseer as Revealed in His Letters*, John Spencer Bassett has amply documented his career.

tion bell, he went to bed only after he had seen that the horses were in the stables, the corn in the cribs, and the Negroes in the cabins. Otherwise, there was danger that supplies might disappear, mules might be ridden, and Negroes might go visiting. If punishment and reprimands were to be distributed, it fell to the lot of the overseer to inflict them. When presents, pardons, and favors were to be granted, the planter and his family gave them to the slaves. The overseer often wanted to attend the rustic dances and merrymakings of his compatriots but his duties kept him on the plantation. No doubt he often slipped away. James K. Polk's overseer writes complaining that the planter does not want him to keep a decent horse or a supply of liquor for himself. Undoubtedly, the overseer was crude, illiterate, and of low social status. At the same time he was often a man of good practical sense and keen business management. His salary usually varied from $250 to $600. He had to remain fairly content in his social and economic status. As a rule he expected to marry his daughter to a man in the like class. But an exceptional overseer who had won the confidence of the factor with whom he dealt might borrow from him, move West, and set up as a planter himself. Not many of them did, it is to be feared. If successful, his descendants by the third generation, having learned the manners of gentility, might be admitted to the society of their peers.

Relations on the plantation have been sentimentalized and moralized over until it is difficult to find an objective presentation. The house servants were the closest to their owner. Their contacts were primary, and their often cordial relations are responsible for the many traditions of the plantation. "The lives of the white and blacks were partly segregate, partly intertwined. If any special link was needed the children supplied it."[26] The field hands lived farther removed, their lives were harder, and the overseer interposed between them and their owners. But unless the master was away most of the time there was likely to be no estrangement. Phillips has phrased it happily ". . . the slaves themselves would not permit indifference even if the master were so inclined. The generality of the Negroes insisted upon possessing and being possessed in a cordial but respectful intimacy."[27]

The connection of the plantation with the economic world outside was usually through the factor. The cotton factor[28] was likely to be a man of money and brains. Since the prevailing prejudice against trade did not operate against him, he was always from a good family. His relations with the planter were personal and intimate as well as economic. Since factorage reflected the hazards of cotton planting, a large profit was allowed him without complaint on the part of the planters. The confidential nature of their relationship attached a high value to the moral hazards. If the planter consigned his cotton to the factor and accepted his accounting usually without quibble, the factor lent the planter money with nothing more definite than a personal pledge. In southern cities such as Charleston, South Carolina, the factors either paid the proceeds of the sea island cotton over to the growers or acted in the capacity of a bank for them, honoring their checks when presented. As a matter of fact, the factor was wholesale merchant, banker, and

[26] Phillips, *op. cit.*, p. 313. [27] *Ibid.*, p. 307.

[28] A. H. Stone, "The Cotton Factorage System for Southern States," *American Historical Review*, XX (April, 1915), 559–66.

cotton agent. He supplied the necessities for the slaves and the comforts and conveniences for the home in jobbers' lots. He handled the planter's money and thus concentrated it in the larger cities. If in some instances he built up a system of rebates on weighing, storing, and drayage, no one blamed him particularly. He was a busy and hard-working man but cotton was the only agricultural commodity that could have stood the delays and exposure at export ports without serious deterioration. If the factors sometimes retired rich men, the hazards of their business were just as likely to wipe out their fortunes over one or two bad seasons. The factors retarded the growth of fair-sized cities in the inland, because they kept out small merchants and country banks. They developed a high standard of business honor, occupied a high social position, and if often angered at seeing their loans for cotton production spent at northern watering places or on European tours, they remained on intimate terms with the planter, directing and advising his undertakings.

The cotton system developed a rationale surprisingly fitted for its self-maintenance. With brilliant, incisive strokes Dodd[29] has analyzed the social philosophy by which the planter aristocracy rationalized their system. It developed the familiar doctrine of inherent inferiority. It taught the degradation of manual labor, the necessity for "mudsills of civilization," and an aristocratic organization of society. "It is the order of nature," wrote President Dew of William and Mary College, "that the being of superior faculties and knowledge, and therefore of superior power, should control and dispose of those

29 *The Cotton Kingdom,* pp. 53, 64–65.

who are inferior. . . . The exclusive owners of property ever have been, ever will and perhaps ever ought to be the virtual rulers of mankind." Admitted that the slaves were inferior by nature there could be no injustice in keeping them so by training. "Odium has been cast upon our legislation," wrote Chancellor Harper of the University of South Carolina, "on account of its forbidding the elements of education to be communicated to slaves. But, in truth, what injury is done them by this? He who works during the day with his hands does not read in the intervals of his leisure for his amusement or the improvement of his mind. If there were any chance of their elevating their rank and condition in society, it might be a matter of hardship that they should be denied those rudiments of knowledge which open the way to further attainments." Finally, Fitzhugh in his *Sociology for the South* propounded that the state must take care that every man, woman, and child shall have a vocation and useful employment with due support. But for the ignorant and poor this demands slavery. "In England the duty of the state is to subordinate the owners of the cotton mills to the government, and the workers should be made slaves of the owners who must give them support and kindly treatment. The American government should grant plantations in the West to responsible men and the landless and idle of the eastern states should be attached to these plantations and become the tenants of their masters for life. Slavery will everywhere be abolished or everywhere be reinstated."

Such was the status of the plantation régime and the human factors in cotton culture in 1860. The cotton system had arrived after a long period of development and change. The plantation in America had begun as a lineal descendent of the feudal organization of agriculture in Europe. It was exported by royal charter, confirmed by proprietary land grants, and accepted by the English gentry as a matter of right and privilege. It was standardized and perfected under tobacco culture. The great excess of land over labor encouraged all attempts to secure labor, and the plantation organized this cheap labor as efficiently as possible around a staple crop. Almost by chance the innovation of human slavery and the invention of the cotton gin in an area of peculiar climatic possibilities for cotton made possible its large scale production. Cotton culture extended the plantation far and wide and settled the Negro in the fertile river bottoms and flood lands where he remains today. It remains to be seen how the plantation survived the shock of the abolition of slavery and, reorganizing itself, fitted into the cotton system of today.

FROM THE COTTON PLANTATION TO COTTON TENANCY

From all accounts the plantation withstood the shock of war and the loss of its great staple remarkably well.[30] Where unaffected by the actual struggle, its organization not only remained intact, but the cultivation of corn, cowpeas, Irish potatoes, sweet potatoes, beans, and fruits in the place of much of the cotton was efficiently carried on. The plantation came to diversify through necessity. More and more the able-bodied white men were drawn away from the great plantation areas, but the system, except where touched by war, went on under

[30] Especially valuable for this period are W. L. Fleming's *Documentary History of Reconstruction,* and *Civil War and Reconstruction in Alabama.*

feminine direction. Fleming says that one could ride for days through the Alabama Black Belt without seeing an able-bodied white man, but the production of the plantation was never higher.[31] The managerial ability of the women of the plantation, the resourcefulness of the Negro artisans, and the faithfulness of slaves in the fields have been deemed equally worthy of comment. The difficulty in feeding the Confederate Army was not one of production but one of commerce and transportation. Sherman's march through the hinterland was well advised as a military measure in that he found an abundance of stores and supplies, products of the plantation, to destroy.

It is useless to ask what would have been the social and economic readjustments in the plantation forced by the abolition of slavery without war and reconstruction. Economic disorganization and social demoralization were bound to occur, but they were doubly accentuated by the shock of war and the deadly conflict of reconstruction. In the combatant areas slavery was destroyed by the friction of the struggle itself, and in the presence of the Union armies the plantation system went down like a house of cards. Its bonds were dissolved; its laborers deserted "when freedom cried out," and became hangers-on around army posts and camp followers of Union armies.[32]

A period of readjustment, a brief hiatus in the cotton system, followed emancipation. The cotton laborer had been given a free status but no higher economic or social standing. He had acquired mobility but no more security.

---

[31] *Civil War and Reconstruction in Alabama,* p. 209.

[32] For description of this period see Walter L. Fleming, *The Freedmen's Savings Bank,* chap. I.

Many Negroes expected the illusionary forty acres and a mule; some met with swindlers who took their money in return for fraudulent deeds. Much has been made of the Negro's refusal to work after emancipation. Some of them no doubt spent their days hanging about the towns to return to their cabins at night. The plantation had inculcated habits of industry and hard work but had linked these responses to the stimulus of personal compulsion rather than economic competition. Such industry was a matter of obedience, not a matter of forethought. Thrift had not been taught because the worker had nothing that he might save. But, as has been suggested, the plantation itself had broken down, and no doubt many laborers did not stay on the plantations simply because their master could not give them work.[33]

The plantations had been returned to the planters, and the high prices for cotton encouraged them to reëstablish its culture. There was a gradual return of laborers to the plantations under the wage contract system. This system met with varying success in different localities, but as a general thing it was abandoned. There were no banks left solvent after the war; the planters were land-poor, and it was almost impossible for them to pay a weekly or monthly wage. Attempts to bind Negroes by contract failed, because they did not understand contracts and refused to wait for yearly wages.[34] In cases of crop or price failure the whole loss thrown on the planter might result in his inability to meet the wages. The superior profitableness of cotton production and the scarcity of labor created higher wages notably

---

[33] See Charles H. Wesley, *Negro Labor in the United States*, chap. V.

[34] M. B. Hammond, *The Cotton Industry*, pp. 123–27.

in the western cotton states. This caused some Negro migration with consequent breaking of contracts.[35] The loss thrown on the planter by crop failures, and the mobility of workers resulted first in a reduction of yearly wages which ranged from $100 to $150 for men in 1867 to a scale of from $90 to $110 in 1868.[36]

The result was the share system, adopted after a period of trial and error, to govern the division of the product between landlord and tenant. The share given the landlord came to range from a half to a "third and fourth." If the worker owns no stock or implements, he lives in the landlord's house, works his farm, pays half of the crop expenses as fertilizer and ginning, and receives half the crop for the labor of himself and family. The term usually applied to such a farmer is cropper rather than tenant, and his status is that of "a servant whose wages depend upon the amount of profit." [37] A tenant has possession in his own right, but the landlord may direct and control the operations of a cropper. The Georgia Supreme Court has declared, "The case of the cropper is rather a mode of paying wages than a tenancy." [38] The share tenant owns his mules and implements of cultivation and pays, for the rent of the land, a third of the corn and a fourth of the cotton. Another system of payment which grew up was called cash or "standing rent." The rental was standing in that a fixed amount was to be paid from the fruits of the soil regardless of how much was produced. This usually means the delivery

[35] T. J. Woofter, Jr., *Negro Migration*, pp. 92–122.

[36] *Report of Department of Agriculture* 1868, cited by Hammond, *op. cit.*, p. 124.

[37] 12 Iredell 3; 123 *North Carolina*, p. 749.

[38] Appling *v.* Odum, 1872, *Georgia Reports*, p. 584–85.

of cotton amounting to a specified cash value at harvest time. Laborers who had worked up to the dignity of standing renters owned their animals and work tools. Since the renters possessed no capital the landlord often found it necessary either to stand for his supplies at the merchant's or to furnish them himself. Thus grew up the crop lien system as will be shown later.

That the system was a successful adjustment for the times was shown by the ease with which it displaced wage labor on the cotton farm. That it deserved to become a permanent arrangement is doubtful. With a measure of truth share renting has been called "a system whereby labor is secured without wages and loans are made without security." The lack of any security except the crop leads in many cases to close supervision over the activities of the renter. How this may fail is told by a Georgia planter:

> The plan of dividing crops under the share system is an equitable one, and if it were properly carried out there could be no cause for complaint; but the owner in nine cases out of ten, has not only to furnish his farm, but to supply all the needs of the tenant, without having any control over the time or acts of the tenant, who is often seen idling and loitering when his crop requires his immediate attention. Tenants owe the owners for provisions, clothing, tobacco, etc., and in many cases they are indifferent as to whether they produce enough to pay the owners these advances made during the season. Thus the landlords annually lose largely by the system of shares, simply because they have all the risks and no corresponding control.[39]

[39] "Report on Cotton Production, Georgia," *Census* 1880, VI, 172.

The changes involved in the transition from slavery to free labor are well shown in the history of a plantation in Middle Georgia:

The plantation underwent all of the changes of the transition period except that in mobility of labor, for none of the Negroes left. In 1860 the Barrow's plantation of a thousand acres at Oglethorpe, Georgia, contained about 25 Negro families living in the quarters centered around the big house. "For several years following emancipation the force of laborers was divided into two squads, the arrangement and method of cultivation being very much the same" as in the ante bellum days. "Each squad was under the control of a foreman who was in the nature of a general of volunteers." . . . "The laborers were paid a portion of the crop as their wages which made them feel interested in it."

After a while, however, even the liberal control of the foremen grew irksome. . . . The two squads split up into smaller and smaller squads, still working for a part of the crop with the owners' teams, until this method of farming came to involve great trouble and loss. The mules were ill-treated, the crop was frequently badly worked, and in many cases not honestly divided. It became necessary to reorganize the plantation. The owner sold his mules to the Negroes on credit, thus placing the risk from careless handling upon the tenants. The gang system was abandoned, and the land was divided so as to give each family its individual tract. When some of them had to walk a mile it became impracticable to keep the cabins grouped. One by one the workers moved their house on to their farms, settling in convenient places near springs. The plantation now contained 999 acres as one acre had been given for a schoolhouse and a church.

The system of sharing was abandoned for cash rent in kind, especially cotton. The Negroes planted what they pleased and worked when they liked, except that the landlord required that enough cotton be planted to pay the rent.

"The usual quantity of land planted is between twenty-five and thirty acres, about half of which is in cotton and the rest in corn and patches. An industrious man will raise three times the amount of his rent cotton, besides making a full supply of corn, sirup, and other provisions." The poorer farmers work "sufficiently well to pay their rent, buy their clothes, spend at Christmas, and let the rainy days of the future take care of themselves."

The following is the yearly budget of one Ben Thomas, his wife, a son and a daughter, one of the best farmers on the Barrow place. He made:

5 Bales of Cotton	2,500 pounds	@	.11	$275.00
Corn	160 bushels	@	.75 per bushel	120.00
Fodder	3,000 pounds	@	1.00 per hun.	30.00
Wheat	30 bushels	@	1.00	30.00
				$455.00

His renting contract included the following:

Cotton	500 pounds	@	$0.11	$ 55.00
Corn	25 bushels	@	1.00	25.00
Fodder	500 pounds	@	1.00 per hun.	5.00
Cotton seeds	40 bushels	@	.50	20.00
				$105.00

Valuing cotton seed at 50 cents a bushel it will be seen that the tenant paid less than one-fourth of his crop as rent. Other tenants not so industrious paid a larger proportion.[40]

The plantation was not undergoing internal change in the midst of a static society. The plantation was re-organizing but it was also breaking up. The neglected and overlooked small farmers of the South were entering

[40] D. C. Barrow, "A Georgia Plantation," *Scribner's Monthly* (April, 1881), pp. 830–36.

the cotton system. The elevation of the unprivileged natives of the uplands to the rank of landowners provided a "remarkable increase in the proportion of whites employed in the cultivation of cotton." [41] Says Henry W. Grady, "the earth hunger of the poorer classes of whites who had been unable under the slaveholding oligarchy to own land was striking. . . . Never perhaps was there a rural movement which was accomplished without revolution or exodus that equalled in extent or swiftness the partition of the plantation of the ex-slaveholder into small farms." [42] The eastern upland areas, inhabited by white farmers, were reclaimed for cotton culture by the use of fertilizers. White farmers moved to comparatively undeveloped areas such as the wire grass in Georgia.[43] Cotton culture was carried to Texas and later to Oklahoma by migrations of white farmers. By 1876 the United States Commissioner of Agriculture stated that nearly 40 per cent of the cotton was grown by white farmers.[44] By 1910, although Negro farmers cultivated 52 per cent of the total cotton acreage in the South, the white farmers produced 67 per cent of the total crop.[45]

The southern small farmers' opportunity to buy land came as a result of falling cotton prices after 1871. Encouraged by thirty-cent cotton, planters had "refitted their quarters, repaired their fences, summoned hundreds of Negro croppers at high prices, and invested lavishly

[41] Hammond, *The Cotton Industry,* p. 129.

[42] "Cotton and Its Kingdom," *Harper's Magazine,* LXVIII (May, 1881), 721–22.

[43] Robert Preston Brooks, *The Agrarian Revolution in Georgia, 1865–1912,* chap. VII.

[44] *Report 1876,* p. 136, cited by Hammond, *op. cit.,* p. 130.

[45] *Negro Population in the United States, 1790–1918,* pp. 594–96. U. S. Census Monograph.

their borrowed capital in what they felt sure was a veritable bonanza."[46] When the cotton market fell, crops produced at high prices on borrowed capital brought ruin in their wake. Lands were thrown on the markets at forced sales, and the market sagged under their weight. "Plantations," writes Grady, "that had brought from $100,000 to $150,000 before the war were sold at $6,000 to $10,000 or hung on the hands of the planter and his factor at any price whatever."[47] Many farmers, both black and white, bought small farms and set up as independent operators. For the first time since the beginning of the plantation régime, land owning bid fair to become generally diffused among the southern people. In the decade from 1860 to 1870 the total value of farms in ten cotton states had declined from almost $1,479,000,000 to $764,000,000, a fall of 48 per cent.[48] While a great part of the loss was due to the ravages of war, part of the loss shows that in times of crop or price failure there are likely to be more sellers than buyers. It was this fall in value that gave an opportunity to the small owners. The following table[49] is arranged to show the break-up of the old régime in the decreasing size of southern farms and farm values as compared with northern farms. In eleven southern states the average size of the farm decreased from 335.4 to 111.4 acres. The greatest single decrease, that from 1860 to 1870, represents the division of plantation into tenant holdings, listed as separate farms by the census.

---

[46] Henry W. Grady, *op. cit.*, p. 722. [47] *Loc. cit.*

[48] Computed from the Census of 1860 and 1870 by Hammond, *op. cit.*, p. 127.

[49] *Census,* V (1910), 878.

TABLE I

COMPARATIVE SIZE AND VALUE OF NORTHERN AND SOUTHERN FARMS 1850–1910

Census Year	Average Acres of Land per Farm		Average Improved Acres per Farm		Average Value of Land and Buildings per Farm	
	The South	The North	The South	The North	The South	The North
1910	114.4	143.0	48.6	100.3	$2,374	$8,182
1900	138.2	133.2	48.1	90.9	1,251	4,190
1890	139.7	123.7	58.8	87.8	1,402	3,721
1880	153.4	114.9	56.2	76.6	1,224	3,314
1870	214.2	117.0	69.2	69.2	1,456	3,463
1860	335.4	126.4	101.3	68.3	3,455	3,180
1850	332.1	127.1	101.1	65.4	2,051	2,380

The movement from landlords to home ownership was checked before completion. In 1881 Henry W. Grady could write:

After sixteen years of trial everything is as yet indeterminate. And whether this staple is cultivated in the South to independence or beggary, are matters yet to be settled. Whether its culture shall result in a host of croppers without money or credit, appealing to the granaries of the West against famine, paying toll to usurers at home and mortgaging their crops to speculators abroad even before it is planted—a planting oligarchy of money-lenders, who have usurped the land through foreclosure, and hold it by the ever growing margin between a grasping lender and an enforced borrower—or a prosperous self-respecting race of small farmers cultivating their own lands, living upon their own resources, controlling their crops until they are sold, and independent alike of usurers and provision brokers—which of these shall be the outcome of cotton cultivation the future must determine.[50]

The future has decided. Cotton has definitely become a tenant crop. More than half of the share tenants in the United States are found in the eight following southern states listed in order: Texas, Georgia, Mississippi, North Carolina, Arkansas, Alabama, South Carolina, and Tennessee. It was shown in the Census of 1900 [51] that of all the farmers to whom cotton was the chief source of income 67.7 per cent were tenants.

A change that had been taking place in the southern landscape brought to the fore new social types, created new economic interests, and made the cotton farm a speculation for urban investors. Towns were springing up all over the Cotton Belt, and in these towns had grown up a hustling, urban type—the rising merchants, lawyers, and doctors. The growth of towns had been hindered before the War by the factors located in the larger cities who had supplied the wholesale needs of the planters and marketed their cotton. With the passing of the factors there grew up the supply merchant who furnished cotton growers on time and bought their cotton. It was this group which secured the passage of the crop lien laws enabling farmers to mortgage in January their as yet unplanted crops for supplies on which to live. The depreciating value of land and the homestead exemptions usual in southern states made merchants unwilling to accept mortgages on land. A conflict of interest ensued between landlord and supply merchant as to whether the

[50] *Op. cit.*, p. 719.

[51] Goldenweiser and Truesdell, *Farm Tenancy in the United States*, p. 83, U. S. Census Monograph IV.

crop lien should be used to secure supplies advanced by the merchant or to secure the rent to the landlord. In many cases the landlord settled the problem by taking the crop lien and standing good for the store bill. In other cases the landlord contented himself with taking a second lien on the crop. As they were in a rather risky business the supply merchants tended to sell their supplies on time at interest rates of approximately 20 per cent. Many of them lost money, no doubt, but it is very likely that the earliest fortunes to grow up in the South out of the dead level of poverty left by reconstruction were made by men who combined the functions of supply merchant and cotton buyer. Said one observer, "the road to wealth in the South, outside of the cities and apart from manufactures, is merchandising."[52]

The supply merchant, forced by the nature of his position to exercise supervision, assumed a paternalistic attitude toward his tenants. "The merchant who has a lien on the tenant's share of the crop," writes an observer, "pays his taxes, buries his wife or child, buys him a mule if he needs one, and feeds and clothes him and his family."[53] The crop lien system opened a way to ownership for the Negroes who started without land or credit. On the other hand, no doubt many of them were more or less victimized. Says W. E. B. Du Bois:

> A thrifty Negro in the hands of well-disposed landlords and honest merchants early became an independent landowner. A shiftless, ignorant Negro, in the hands of unscrupulous landlords or Shylocks, became something worse

---

[52] G. K. Holmes, "The Peons of the South," *Annals of the American Academy,* IV (Sept., 1893), 265–74.

[53] *Loc. cit.*

than a slave. The masses of Negroes between the two extremes fared as chance and the weather let them.[54]

"The Negro," said a southern governor with the wasteful methods of tenant cultivation in mind, "skins the land and the landlord skins the Negro."

As planters failed and plantations went on the block the rising urban groups tended to step into the class of the landed gentry. Merchants, lawyers, and doctors bought plantations as speculative investments. As the urban class began to buy farms for investment, the price of land tended to rise beyond reach of the small cotton grower. "There is," wrote Grady, "a sure though gradual rebunching of the smaller farms into large estates and a tendency toward the establishment of a landholding oligarchy." [55] Many landlords also moved to town. Absentee landlordism, too, brought lower yields, less supervision and contact with tenants, and a lower standard of living for them. Farming became a financial instead of an agricultural interest. Cotton, the money crop, came to be exacted by landlords living in town, and the tenant's interest in diversification and food supplies was disregarded.

The rise in tenancy in the South is not to be denied. The percentage of tenancy for the United States is 38; in the South there are eight states with over 50 per cent tenancy. No figures on tenancy are compiled for the United States before 1880 but since then the percentage has steadily increased for all the cotton states. The states that had 35 to 45 per cent tenancy when Henry W. Grady wrote now have over 60 per cent:

---

[54] *Negro Landholders in Georgia,* U. S. Dept. of Labor Bulletin 35.

[55] "Cotton and Its Kingdom," *Harper's Magazine,* LXIII, 734.

TABLE II

INCREASE IN PERCENTAGE OF TENANCY IN COTTON STATES—1880–1925

	NORTH CAROLINA	SOUTH CAROLINA	GEORGIA	ALABAMA	MISSISSIPPI
1925	45.0	65.1	63.8	60.7	68.3
1920	43.5	64.5	66.6	57.9	66.1
1910	42.3	63.0	65.6	60.2	66.1
1900	41.4	61.1	59.9	57.7	62.4
1890	34.1	55.3	53.5	48.6	52.8
1880	33.5	50.3	44.9	46.8	43.8
	TENNESSEE	ARKANSAS	LOUISIANA	OKLAHOMA	TEXAS
1925	41.0	56.7	60.1	58.6	60.4
1920	41.1	51.3	57.1	51.0	53.3
1910	41.1	50.0	55.3	54.8	52.6
1900	40.6	45.4	58.0	43.8	49.7
1890	30.8	32.1	44.4	0.7	41.9
1880	34.5	30.9	35.2	—	37.6

In cotton growing areas in ten southern states, according to 1920 census figures, there were 1,030,321 farm owners and managers to 1,345,805 tenants, a ratio of 45 owners to 55 tenants. Of these renters 284,180 were classified as cash tenants, 487,770 as croppers, and 550,879 as share tenants or third and fourth renters. Thus out of a hundred cotton renters approximately 21 are cash renters, 37 are croppers and 42 are share tenants. They are distributed by states as follows:

TABLE III

FORMS OF TENURE IN COTTON STATES—1920

State	Owners and Managers	Tenants					Per Cent of Total	
		Total	Cash[1]	Share[2]	Cropper	Un-specified	Owners and Managers	Tenants
North Carolina	152,304	117,459	15,916	59,287	39,939	2,317	56.5	43.5
South Carolina	68,462	124,231	40,072	38,944	43,789	1,426	35.5	64.5
Georgia	103,778	206,954	61,820	45,390	97,497	2,247	33.4	66.6
Alabama	107,830	148,269	57,694	37,209	47,897	2,469	42.1	57.9
Mississippi	92,299	179,802	38,972	51,608	86,859	2,363	33.9	66.1
Louisiana	58,082	77,381	13,505	31,692	31,309	875	42.9	57.1
Arkansas	113,382	119,221	18,171	51,056	47,665	2,329	48.7	51.3
Tennessee[3]	36,308	42,343	9,907	15,744	15,508	1,184	46.2	53.8
Oklahoma	94,152	97,836	15,710	72,340	8,926	860	40.0	51.0
Texas	203,724	232,309	12,413	147,609	68,381	3,906	46.7	53.3

[1]Includes standing renters.
[2]Includes share-cash tenants.
[3]Figures represent only 22 counties in West Tennessee.

These figures have varying interpretations. The great increase of tenancy in Texas and Oklahoma is accompanied by the introduction of cotton culture but lacks altogether the historical explanation of the plantation. On the other hand, it is due to the peculiar development of the states from another capitalistic form of land utilization. Not the break-up of plantations but the decadence of cattle ranches accounts for the increasing tenancy in this new area. Speculative methods of acquiring land made it possible for settlers to secure large tracts. On the change from a ranching economy the owners let parcels of these farms to immigrants for the cultivation of cotton. As a result, Oklahoma within the comparatively short time of thirty-five years has increased her tenancy rate from 0.7 to 58 per cent.

On the other hand, in plantation states an increase in the percentage of tenancy is likely to be an indication that Negro wage hands are rising into cropping and share renting. The decrease in percentage of tenancy in Georgia from 1920-1925 is not to be taken as representing the rise of any group into land ownership. On the contrary, it is due to the migration of tenants from the plantation areas. While it has played a part, it is impossible at our present state of knowledge to know to what extent increasing tenancy in the South represents investment buying of land, adding of field to field in the creation of new plantations.

THE PLANTATION TODAY

That the cotton plantation is a real factor in the present cotton system is well known. As already shown, the change from slave to free labor [56] involved the substitution of cultivation by tracts or parcels of land instead of cultivation by the whole plantation as in the gang system; it gave the laborer mobility, legal freedom of contract, and wages, or a share of the crop in return for his work. But the outstanding feature of the old plantation was supervision, and it remains characteristic of the new. This control, vested in a unified administration, extends to choice of crops, distribution of labor, and marketing. To speak of southern and northern tenant farms in the same terms is an anomaly. Tenant farms in the Middle West are usually as large as owner farms, are operated independently by a farmer who employs hired labor and possesses as high a standard of living as his landlords. The census has failed to show the prevalence of the plantation, for it has counted each tenant's or cropper's holding as a separate farm. The presence of these plantation strips was shown in the census by the contrast between northern and southern farms. In 1910 the improved acreage of the average southern farm was 48.6, its value $2,374; the improved acreage for a northern farm was 100.3 and its value was $8,182.

The special census of plantations [57] in 1910 showed the extent of that form of land tenure. There were tabulated 39,073 plantations, containing five or more tenants on 28,296,815 acres of farm land, of which 15,836,363 acres were in crops. These plantations, as suggested in Chapter II, were localized in areas of fertile land suitable for cotton production and coincided with the black belts as

[56] See also Phillips, "The Plantation with Slave Labor and Free," *American Historical Review*, CCXIX (July, 1925), 21–29.

[57] "Plantations in the South," 1910 *Census*, V, 877–90. For census definition of the plantation see Chap. II, p. 31, of the present volume.

localized by slavery. In the Coastal and Piedmont regions of the Atlantic states, twenty-one counties in North Carolina, thirty-five in South Carolina, and ten in Georgia were found to have approximately 15 to 30 per cent of their improved land in plantations. In Alabama forty-seven counties mostly in the Black Prairies had around 30 per cent of their improved land in plantations. The Mississippi River Valley had around 45 per cent of its improved land in plantations in forty-five counties in Mississippi, twenty-three in Arkansas, eleven in southern Tennessee, and twenty-nine in Louisiana. Slavery had carried the Negro here, and mosquitoes, malaria, and the high cost of the rich river bottom land have kept white farmers out. In and near the Black Prairie of Texas, forty-one counties were found to have about 15 per cent of the lands in white tenant farms comparable to the plantation system.[58]

The average plantation was found to contain 724 acres of which 405 were in improved land. The value of its land and buildings was $17,322. Its acreage thus was more than five times and its value three times as great as those of the average farm in the United States. The average farm retained by the landlord for cultivation was worth $6,564 and contained 330 acres, 26 per cent of which was improved. There were 398,905 tenant tracts which covered 15,367,398 acres of farm land. The average tenant farm contained 38.5 acres of which 81 per cent was improved. Although composed of richer and better cleared land the value of land and buildings was

---

58 1910 *Census*, V, p. 880. See also Dept. of Agriculture Bulletin 1269, Fig. 2, p. 5. The plantation was noticeably absent in counties where soil and topography rendered impossible the production of cotton.

less than $1,000.00. The average value of the buildings was $179 per tenant farm. Each plantation averaged over ten tenant farms.[59]

Three types of plantations are discernible in the Cotton Belt: the small plantation, managed by the landlord who lives on it and directs to some extent the work; the large plantation, owned by a well-to-do capitalist or possibly a corporation and run by a managerial staff; and the plantation bought as a speculation by one engaged in another business who attempts to operate it as an absentee landlord.[60] The management of a plantation owned by a corporation is likely to be most efficient; the smaller plantation on which the landlord lives usually possesses a well-developed social system as well as economic organization. The plantation bought as a speculative venture and owned by an absentee landlord is likely to be least efficiently managed and to offer its tenants a lower economic status.

The system of management to be described is followed in the main by all plantations but is best developed in those owned by larger corporations. The owner or general manager has the difficult task of supervision. It is the owner's task to look after the financing of the plantation by keeping up connections with credit institutions, to look after purchasing supplies, marketing products, and with the aid of a bookkeeper to oversee the accounts of all the different tenants. The man next in authority is called the farm manager. The successor to the ante bellum overseer has risen in intelligence, social status, and managerial ability. He directs the planting and cultivating of crops

59 1910 *Census*, V, 881.

60 Goldberger, *Pellagra in the Mississippi Flood Area*, Public Health Reprint 1187, pp. 10–11.

and the operation of the gin and supply store. A gin mechanic is kept only during the active ginning season, but if the plantation is fairly large a store manager is employed the year around. It is the business of the farm manager to look after the work of tenant farmers and the wage hands whom he delegates to the charge of an overseer. In addition the hired white labor in care of stock, blacksmith shops, and grist mills must be apportioned.[61]

The modern cotton plantation is a unified organization, but because of the very nature of agriculture there are many difficulties in the way of centralized supervision. The work, stock, tools, and the crop, whether belonging to landlord or tenant, are looked after. In order to avoid loss, the manager supervises the tenant's financial dealing so that his living expenses advanced do not exceed his productive capacity. Tenants may desert at critical times, leaving debts unpaid and crops untended. These contingencies require in the plantation manager a skillful blend of tact and firmness in dealing with the human elements on the plantation.[62] Men of such qualities are hard to find and when found often work up to other business opportunities or become plantation owners themselves. The average salary for plantation managers in 1920 was $2,100 a year and included free house rent, food for the family, and privileges of pasture for live stock.[63] Assistant managers and overseers received $1,550 and $1,000 with perquisites. The average cost of manage-

---

[61] Brannen, *Relation of Land Tenure to Plantation Organization,* Dept. of Agriculture Bulletin 1269, p. 12.

[62] See *ibid.,* p. 11 for a discussion of this topic.

[63] *Ibid.,* pp. 16–17.

ment on cotton plantations was approximately $1.83 per acre in crops.

The labor on plantations is divided between wage hands, croppers, and share tenants. The terms as to tenure are the same as those on rented farms in the South not in the plantation area but, as suggested, the workers are more closely supervised. A greater per cent of the workers are wage hands than in other tenancy areas. At least 25 per cent of the work is done by utility hands who feed stock, dig ditches, and do carpenter work, extra day hands called in for the emergencies of chopping and picking, and regular hired hands who raise feedstuff for live stock and work the landlord's crop through and through on the old gang system. In 1913 such hands received $18 to $25 per month without board or $12 to $16 per month with board and rations. By 1920 wages had increased almost 100 per cent. Day hands who had received $1.25 per day received $3.00. The landlord is relieved of the risk of advancing supplies to the wage hand for he pays as he goes. The monthly rations resemble those of the old plantation: 16 pounds of salt pork, a bushel of meal, and one or two gallons of molasses.[64]

The position of the cropper is about half way between that of a tenant and a hired servant. The landlord furnishes the land, a house, tools, and work animals to the cropper who furnishes the labor for half the crop. Expenses of fertilizer and ginning are usually divided equally. In addition, the plantation owner furnishes supplies for which he takes a lien on the crop. If the cropper needs extra money he may work on the plantation as a

---

[64] For discussion see *ibid.*, pp. 19–27.

day laborer, and his acreage is sometimes kept low in order that he may work the landlord's crop.

After the wage laborer has advanced to the cropper stage, he may save money and become a share tenant. He reaches this stage by investing in his own work animals and implements of cultivation. He still receives his house and land from the plantation and pays usually "a third and a fourth" of corn and cotton respectively, as rent. "Standing rent" of so many bales of cotton regardless of the crop has developed in regions like Georgia, where crop failures are not so common. On an average, the share tenants make up about 54 per cent of all renters on plantations.[65] It is the tendency in good cotton years for the plantation workers to advance in status; wage hands become croppers, and croppers become share tenants. It is virtually impossible, because of the high price of land, for Negroes to work up into land ownership in a plantation area such as the Delta. Regions affected by the Negro migration, however, have seen competition for the services of laborers result in a general raising of the tenure status of the hands and croppers.

The following description is of a plantation of 4,200 acres with 130 Negro families in one of the Delta counties of Mississippi:

> Although the acreage in the holding was contiguous, it was considered large enough to handle as two quite distinct plantations, the owner hiring an overseer for each. The landowner owned and operated the cotton gin, the store, and the sawmill that served both the plantations. A considerable acreage was operated by wage labour, the owner hiring labour to farm all of the 222 acres in crops cut for hay, 283 acres, or about three-eighths of the acreage, in maize, and 108 acres, or about

---

[65] *Ibid.*, pp. 29–38.

a fourteenth of the acreage, in cotton. Wage labour was also hired to feed swine, raised to provide part of the pork sold to the negroes, and to care for the mules, kept to farm the land operated by the wage hands and by the share hands or croppers.

Each of the 130 negro families was provided with a cabin. Less than half of these negro families owned mules and farming implements. The negroes who owned the mules and implements with which they worked, farmed only about 40 per cent of the entire cotton acreage and about 30 per cent of the entire corn acreage of the two plantations. Altogether they had 45 of the 130 mules on the entire plantation. The average negro family had about 11 acres in cotton and 4 acres in corn, the basis of assigning acreage being to allow about six acres of cotton and two acres in corn for each male adult worker or equivalent in children able to work. Advances were made to the negroes from March first on. These were held down as near as possible to supplies in value not exceeding half a dollar per acre per month, until August first. From then on only half that amount was allowed. Tenants in need of more advances were given a chance to work for wages part time. The mules used by the croppers were kept in central barns and cared for and used in part by labourers working on a wage basis. These croppers were also furnished with implements. They got half of the cotton and corn raised, for furnishing the labour and for meeting half of the expense of ginning. Advances, made prior to settlement, were deducted from the cropper's half of the receipts from cotton.

The landowner preferred croppers working on the halves to renters owning their own mules and tools, yet, in order to get enough satisfactory labour both kinds of tenants were employed. Renters with their own operating capital furnished the feed of the work animals, the seed, and labour and gave as rent a fourth of the cotton and a third of the corn. They are called "third and fourth renters." The rent paid on the corn land had to be at least ten bushels per acre, this provision being made to discourage corn growing on the part of shiftless renters. Third and fourth renters were given advances, if they had need of them, and these advances were withheld by the landlord from the tenant's share before settlement was finally made. If the mules owned by these renters became sick or died, they were allowed to rent mules owned by the landlord if unable to secure others when needed.[66]

The account of a Negro cropper on a large corporation plantation during an exceptionally successful year is here analyzed:

In 1921 T. J. Weeks, Sr., cultivated on the halves thirty acres in cotton and two in corn for the Eutaw unit of the great Scott Plantation at Scott, Mississippi. The working force of this family consisted of five males and three females. One small child was classified as a non-worker. Weeks must have been an exceptional farmer for he had saved enough from his Christmas so that he did not begin to draw on the commissary until May 10. Between May the 10 and the date of settlement, December 14, he "traded $367 worth." One hundred and fifty-two dollars and seventy cents were for expenses of the crop distributed as follows:

For breaking and planting	$60.00
For three hoes	3.00
For cotton sacking	7.70
For boll weevil poison	60.00
For fertilizer	22.00

He bought a cow and a calf for $50.00, and spent $134.80 for merchandise and rations in addition to what he grew at home. This included $5.40 in cash. In addition, there was

[66] H. A. Turner, "The Share Renting of Farms in the United States," *International Review of Agricultural Economics* (Oct.-Dec., 1923), pp. 518–19.

illness in the family, and he paid $13.00 in doctor bills and a $15.00 hospital fee to the plantation hospital. He proved that he was an agriculturist by paying $1.50 for a subscription to the *Cotton Farmer.*

He produced twenty-four bales of cotton, which sold for 25 and 26 cents, and 21,550 pounds of seed, which sold at $45.00 a ton. In all, Week's crop of cotton brought $3,251.57.

From this was deducted $39.60 levee tax, $152.40 for ginning charges, and $12.00 for planting, leaving his half of the crop $1,517.78. From this was deducted his store account of $367.00 and $26.00 rent for the two acres of corn, leaving the family of eight a balance of $1,124.78 in cash for the year's work.[67]

"The overseer of the old régime has been replaced by the farm manager in the new order,"[68] but supervision is none the less present. It is usually understood in advance that the landlord or manager is to direct the tenant's farming activities. On many well-supervised plantations the farm manager makes daily rounds of inspection giving instructions on details of field work. On 93 of 144 plantations the bell rings about sunrise, and the work day is from sun to sun. Workers who fail to respond to the bell are questioned and reprimanded. The working week usually consists of five and a half days, Saturday afternoon being taken off for shopping. Much care is taken to see that mules belonging to the plantation are not abused, many planters fearing that the Negroes will ride the animals at night and on Sundays. Tools and implements are lent for the year and returned to the plantation stock at the end of the season.

---

[67] Facsimile from *Cotton Estates Incorporated,* by Theodore Price, pp. 56, 57. Copy in Library of Congress.

[68] Brannen, *op. cit.,* pp. 29, 42–52.

Like its ante bellum prototype the plantation has developed a social system commensurate with its economic organization. In some cases community and family affairs are subject to review by plantation managers. Many operators do not allow tenant families living without legal marital relations on their plantations. The landlords determine all the holidays that shall be observed. Plantation barbecues and picnics are given by the management for all the workers.

Plantations on which the owner and his family live have a strongly paternalistic régime. A not too friendly critic of the modern plantation and its human factors writes:

> The patience with which a Mississippi planter deals with his dull, irresponsible labor is almost unbelievable to a Northerner. If he is harsh in punishing a "runaway," too shrewd in his contracts, quick to take advantage of all his opportunities to exploit, the planter is also the long suffering guardian of difficult children. If a hand falls sick, the planter's physician is called. The planter purchases the necessary medicines, and he or his wife watches through the night beside the sick person, for no plantation Negro can be depended upon to administer medicine regularly. Family or community quarrels are patiently heard and decided. Many planters give an annual barbecue where all hands are invited to a feast and merry-making. A Negro cheated by another Negro or by a white man can count on his "boss" to safeguard his rights. The planter protects his Negroes from the countless "agents" who are always trying to sell the hands some trifle at an exorbitant price. The average planter looks upon his hands as responsibilities to be fed, clothed, guarded, and cared for in sickness or disaster. At the same time, he is unalterably opposed to anything that would help these children grow up.[69]

The presence of the plantation system is a unique ex-

---

69 Beulah A. Ratliff, "Mississippi," in *These United States,* I, 32.

ample of an adjustment of man to land and of race to race accepted from history as a social heritage. The plantation possesses an influence in southern life greater than its extent. It tends to set the mode by which the human factors in cotton shall be regulated. In times of prosperity investment in farms as plantations has tended to draw more workers into tenancy. That plantations should come to be vested more and more in the control of corporations would not surprise one student of history. "Its concentration of labor under skilled management," writes U. B. Phillips, "made the plantation system with its overseers, foremen, blacksmiths, carpenters, hostlers, cooks, nurses, plow-hands, and hoe-hands practically the factory system applied to agriculture." [70] But more than any factory the plantation is devoted to the production of a staple from which it cannot escape. It is this devotion to cotton which made the plantation in the beginning, caused its partial break-up after the Civil War, and holds it in hazard today. For the plantation cannot diversify; it never has diversified but once and then under the stress of war. It must have its staple cash crop, and it must produce that crop regardless of the effect on the market. Herein lies the second fact, the menace of the plantation to the whole cotton system. It does not diversify, it does not reduce acreage. The fact that it may produce cotton efficiently does not alter the charge that it produces too much cotton. By depressing cotton prices it not only threatens the human factors in the whole cotton system, it threatens its own existence. The plantation even more than the average farmer has all its eggs in one basket. And that basket is subject to the hazards of the weather, the weevil, and the market.

[70] "Decadence of the Plantation," *Annals of the American Academy*, XXXV (1910), 37.

CHAPTER IV

THE RISKS OF COTTON PRODUCTION: THE WEATHER AND THE WEEVIL

It is a more or less common saying throughout the South that cotton is dynamite. The phrase means that the two million and more of southern farmers who with their families owe their living to cotton depend upon a commodity subject to the widest variations from year to year. From his first apprenticeship the cotton farmer learns to expect the unexpected. This lesson, however, has not taught him how to provide against it. In fact it is doubtful if cotton farmers can provide against their two greatest risks: the hazard of crop failures, and the hazard of fluctuating prices. Cotton has been and is studied every day from the standpoint of the consumer. Statistics are gathered and interpreted for spinners, and much concern is expressed over shortages in world supply. Few serious studies have been made, however, of the producer as he encounters the vicissitudes of failure, near-failure, and bumper crops.

The averages of the risks of production for the whole Cotton Belt are set forth in the statistics of acreage planted, acreage harvested, and production in standard bales of five hundred pounds. From these averages the Department of Agriculture has worked out the average number of pounds of lint produced per acre. That the risks of production are very real things even when distributed by the averages over the whole belt is shown by the fact that as late as 1921 the production of cotton was less than eight million bales—the lowest crop since 1895. This disastrous crop was not caused by restriction of acreage. The yield per cotton acre for 1921 fell to 124.5 pounds of lint, the lowest of which the Department has any record.

The average yield per acre has varied from this low average of 124.5 to as high as 221 and 222 pounds in 1898 and 1897. An examination of Table VI on p. 125 will show distinct trends in the yield per acre. Up to 1890 the general trend was downward; from 1890 to 1905 the trend was upward, after which it was downward until 1909. From 1909 the trend of yields rose to the high level of 209.2 pounds per acre in 1914. The trend since then has been steadily downward with the exception of a fair yield in 1920 of 178.4 pounds.[1] It is yet too early to predict whether the movement upward from the low level of 1921 is of significance. The record crop of over sixteen million bales in 1925–26 was mainly owing to immense increase in western acreage as the yield was only 167.2 pounds per acre. The year 1926 saw an average yield of 182.6 pounds of lint per acre, the highest since 1914. In 1927 this had slumped to 152.3 pounds.

Four main factors operate to cause changes in the yields per acre: shifts in areas of cotton culture, use of fertilizer, weather conditions, and the depredations of insects, chief of sinners being the boll weevil.[2] Of these four factors only the weather conditions and insect damage may be counted as hazards of production. It happens, however, that they are the chief factors influencing yields.

---

[1] "The Cotton Situation," Dept. of Agriculture *Yearbook*, 1921, p. 336.

[2] *Loc. cit.*

A uniform downward trend over a period of time is not likely to be due to either the weather or the weevil, but more often is due to the exploiting of low-yielding areas. Such sections are brought into cultivation as higher prices make it profitable. Although the margin of profit may be lower, the cultivation of low-yielding land cannot be regarded as an unforeseen hazard of production. The upward tendency in yields can be explained, apart from hazards, by the increasing use of fertilizer in the Eastern Belt. The use of fertilizer adds to the cost of production but increases the margin of profit by increasing the yield.

An average of the ten-year trends in yield per acre by states from 1866 to 1915 illustrates some of these general tendencies.[3] The average yield per acre is lowest in Alabama, due possibly to the inefficiency of Black Belt cultivation, and in Florida, whose sandy soils are poorly suited to cotton culture. During the last three decades the increasing yield per acre in the eastern states can be attributed to the growth in popularity of fertilizers. The highest yields are in Missouri and North Carolina, both until recently, out of the main path of the boll weevil. Up until 1921 the averages of yield in the Western Belt and Alluvial Valley declined because of the weevil. In Texas and Oklahoma an added factor was the extension of cultivation into regions of low rainfall.

Frank H. Vanderlip, writing in 1916, gave four-tenths of a bale an acre as our average production of cotton and compared it with 546 pounds of lint per acre grown under experimental conditions. He cited this low yield as "one of the most gigantic examples of incompetency to be

[3] *Cotton Atlas,* Fig. 45, p. 14.

found anywhere."[4] Such a statement fails to take account of the hazards of cotton growing in that it charges the accidents by flood and field and weevil, which the averages distribute, all to one factor, the inefficiency of cultivation.

The Department of Agriculture, Bureau of Crop Reports, estimates the condition of crops on a basis of 100, representing a normal condition. "This normal is neither an average or ideal crop, but what the fields ought to produce with the normal modes of farming, with normal weather conditions and without unusual loss from disease, insects or other adverse influences. The yield per acre under such favorable, though not extraordinary conditions, would be a normal yield, which is more than an average yield but less than a maximum possible yield. A normal yield for one farm or section may vary widely from that of another."[5]

From this basis of a normal yield the crop reporters send in estimates of loss resulting from various causes. The losses from causes over which man has no control such as weather and weevil (until the discovery of methods of poisoning) have within the last fifteen years ranged from 26 to 52 per cent of the "normal crop."[6]

Such losses in many cases may safely be counted as hazards of production or in the good old legal phrase "acts of God," without reflection on the industry or efficiency of the planter. Their reactions on the economic basis of southern society have been serious and profound.

---

[4] Frank H. Vanderlip, *Our Inefficient Acres*, p. 5. Pamphlet.

[5] *Monthly Crop Report*, Aug., 1918.

[6] Dept. of Agriculture *Yearbook*, 1925, Table 327, p. 956.

THE WEATHER

Until the appearance of the boll weevil the weather figured as the chief factor of the risks of nature likely to make or break the cotton farmer. Most planters would agree, no doubt, with the statement of one that "the cotton plant is the most tender plant that grows. It is susceptible to more disasters and requires more of human manual labor to produce it than anything else."[7]

On the other hand, the disasters that occur to cotton may be regarded as due to the unusually severe conditions which it has to face. An expert of the United States Weather Bureau writes:

> Cotton, though a sensitive plant, is of all summer-growing crops of the South about the least affected by ordinary changes in the weather. Its long period of growth, fruiting, and maturity affords it ample opportunity to recover from a number of temporary set-backs. During the protracted season from planting in April to the completion of the harvest in November, it is exposed to many varieties of weather and it seems to endure the bad as well as enjoy the good.[8]

Owing either to the delicacy of the plant or the vagaries of the American climate, reductions from full yield per acre due to weather have varied in the period from 1909 to 1925 from 13 to 29 per cent.

The most favorable weather conditions for cotton have been thus summarized by J. R. Marbury, meteorologist:

---

[7] Representative Burleson of Texas at a hearing before the House Committee on Agriculture, Dec. 19, 1905.

[8] J. R. Marbury, "Relation of Weather Conditions to Growth and Development of Cotton," Dept. of Agriculture *Yearbook*, 1904, p. 141.

April. Frequent but light showers to keep soil moist for germinating seed. Too much moisture tends to rot the seed and cold rain prevents sprouting.

May requires but a small amount of rain in order to assist the farmer in cleaning out weeds early.

June to the middle of August is the growing period. June calls for plenty of sunshine, very little rain with air dry during the day but dew at night.

July to the middle of August the crop can stand some rain if the season has been good.

August. For the blooming period plenty of warm sunshine with light rains about ten days out of the month. The normal weather conditions of the Cotton Belt are admirably suited to the blooming seasons. Too much rain at this time, however, causes overgrowth of the weed and the squares may drop off.

September and October require dry weather for the picking season. Rains cause decay of the bolls and serve to discolor and beat out the fiber.[9]

Droughts are more frequent and of greater destructiveness in the Western Belt, causing the greatest disaster in western Texas, not so much in Oklahoma, and less in Arkansas and west Louisiana.[10] Lack of moisture causes the growth of the plant to be stunted. One or two small bolls bloom at the top and the growth of the plant is over unless it rains. If rain falls the plants start up again and in six weeks will produce a new growth of bolls. From this possibility, Hubbard says, comes the saying that "Texas can promise more and make less and promise less and make more than any other cotton state."[11]

---

[9] *Ibid.*, pp. 141–45.

[10] Hubbard, *Cotton and the Cotton Market*, p. 54.

[11] *Ibid.*, p. 35.

Also peculiar to the West are the hot winds which become devastating enough to wilt corn in a day, turning the leaves brown. Cotton offers more resistance but wilts after a time under such onslaughts. In 1909 hot winds are credited with reducing the yield 3 per cent of the average of the whole belt, which means that the localized damage must have been very great.

Important in connection with the late crops induced by drought is the date of the first killing frost. Frost destroys the late maturing bolls and thus may cause a serious loss of part of the crop. The earliest frost recorded in the Cotton Belt was noted at Memphis October 9 and reached to the Gulf.[12] The frost damage that year is estimated at 6 per cent. As a usual thing it ranks less than 1 per cent. In addition to threatening western crops delayed by drought, frost is apt to be severe in the part of the Piedmont Plateau nearer the mountains.

Measured by their damage to the cotton yield, excessive rains rank next to droughts as a climatic risk of production. The damage has ranked as high as 15.3 per cent of a normal yield in 1919, but the average reduction is between 5 and 6 per cent of the yield. The Eastern and Alluvial Belts are much more apt to suffer from too much rain than Texas and Oklahoma.

The rains that cause the greatest losses in cotton production come at two periods: the early rain in April, May, and June, and the wet falls. A wet April, it was pointed out, delays planting and causes seed to rot in the ground; a wet May causes the plant to shoot up in weed without sending down an adequate taproot. Consequently, when the long, hot, dry days of midsummer

---

[12] *Ibid.*, p. 36.

come the plant cannot extract moisture from the subsoil and sheds its bolls. Since the advent of the boll weevil the wet spring has become even more dangerous to the cotton planter.[13] The heaviest spring rainfall occurs in the Central Belt. The heaviest August rainfall is along the Eastern Gulf Section and the South Atlantic Coast. Wet falls are, as Hubbard points out, the bane of the cotton buyer. Even if the fiber is not discolored or beaten out, cotton may be picked wet, or the damp, half-rotted bolls may have seed sprouting in them. In either case the grade is lowered, and the staple rendered unsatisfactory to the spinner. The farmer gets much less for his product and in many cases finds he would have saved money by foregoing the expense of picking a damp crop. The years 1906, 1913, 1919, and 1925 are examples of wet falls that gave much trouble.[14]

COTTON DISEASES

Diseases have never ranked high as a risk of production of the cotton crop. In 1912, however, a 4.3 per cent reduction was estimated as caused by cotton diseases. In 1920 a plant disease survey of the Department made an estimate of total losses of cotton from disease at 13 per cent,[15] possibly an overestimate. Several of the diseases are due to fungi. In cotton wilt the parasites enter the roots and plug the water vessels, causing wilting and death of the whole plant. Only resistant types of cotton can be grown in affected areas as the fungus remains in the soil indefinitely. Another fungus growth causes the Texas root rot peculiar to soil of the type of the black

---

[13] *Ibid.*, p. 30.

[14] *Ibid.*, p. 37.

[15] "The Cotton Situation," p. 355.

waxy. Anthracnose is caused by a fungus which produces boll rot in wet weather. Root knot is a swelling of the roots of the plant caused by the presence of a small ell worm. The bacterial blight known as leaf spot although present in most cotton fields does not seriously affect upland cotton. Rust is a physiological disease of the plant, caused by lack of potash and vegetable matter, or by poor drainage.[16]

INSECT PESTS

Several insect pests that prey upon the cotton plant have been long known and feared. Farmers have learned in the early fall to look for the "army worms" which suddenly appear in great numbers, march across the fields, and leave the plants bare of foliage. Their sudden appearance is owing to the fact that the adult moths fly from the tropic regions to deposit their eggs in the Cotton Belt. As the cotton leaf worm does not attack the bolls its damages are excessive only when the leaves are stripped early enough in the season to affect the growth of the plant. Cold weather finishes the pest until the next season. Calcium arsenate is a sure poison.

The cotton bollworm is a general feeder known for its presence in sweet corn. At one time its damage to the cotton crop was estimated at $8,500,000 yearly. Its ravages are sporadic, however, and calcium arsenate is an effective poison. Cutworm, plant lice, and red spiders are minor insect pests.[17]

Possibly the most dangerous single enemy to the cotton plant is the pink bollworm, first noted in Egypt in

[16] *The Cotton Plant,* Dept. of Agriculture Experiment Station Bulletin 33, pp. 271–317.

[17] *Ibid.,* pp. 317–51. Also, "The Cotton Situation," pp. 352–55.

1911. By 1916, Department entomologists knew of its appearance in Mexico. It appeared in Texas between 1917–1921 and was apparently stamped out by severe measures. Particularly dangerous is the fact that the larvæ enter the cotton seed and remain there for several months. In February, 1928, a new and extensive outbreak was reported in Ector, Midland, and Martin counties, western Texas.[18]

THE BOLL WEEVIL

Today the world's largest consumer of raw cotton is the boll weevil. None of the risks attendant upon cotton have equaled this little insect in the drama of its appearance and the tragedy of its effect. In 1896 Dr. L. O. Howard, Entomologist of the Department of Agriculture wrote:

> . . . the cotton plant cannot be said to suffer seriously from the attacks of insects. . . . An exception to this general statement may in the future be found in *Anthonomus grandis,* a Mexican weevil which damages cotton bolls. This insect down to the close of the season of 1894 was known to us only through a few specimens collected upon cotton bolls in Mexico some years earlier by Dr. Edward Palmer. During 1894, however, we learned that the species had made its appearance in the state of Texas.[19]

It had crossed the Rio Grande near Brownsville about 1892. The Mexicans called the insect the "picudo," referring to its long snout. The Texas planters called them "sharpshooters" because of their ability in puncturing the bolls. Dr. Howard wrote as early as two years after

[18] *New York Times,* Feb. 15, 1928.
[19] *The Cotton Plant,* pp. 319–20.

the first weevils were found: "There can be no doubt that this insect may become the most serious enemy to the cotton plant with which cotton growers in this country have had to contend and every effort should be made to prevent its further spread,"[20] and added "none of the general applications of insecticides will be of the slightest value against this species."[21] From the time the weevil first invaded Mexico in 1848 no organized fight had been made against it. As a result, large areas were laid waste and taken out of the area of cotton culture.

Texas followed the same course of non-resistance. Dr. Howard in 1894 advised the passage of legislation empowering agricultural commissioners of counties to enforce remedial measures. A report of Professor Townsend of the Division of Entomology in 1895 earnestly advised the creation of a cotton free zone along the Rio Grande. The recommendation was not taken seriously. Texas failed to realize the extent of the peril to the cotton industry. Added to this was a natural hesitancy in forbidding to the farmers who always raised the first bale in the United States the chief crop which the climate permitted.[22] It has been said that in the early days a ten-mile no-cotton zone properly enforced would have stamped out the evil.[23] Such measures were taken in 1917 against the pink bollworm after Texas had learned her lesson. This insect, which is estimated to destroy 50 per cent of the crop, was found in a cotton field at Hearne, Texas. The Hearne district was then made a

---

[20] *Ibid.*, p. 342. [21] *Ibid.*, p. 339.

[22] W. D. Hunter, "Present Status of Boll Weevil in the United States," Dept. of Agriculture *Yearbook*, 1901, p. 371.

[23] C. M. Munn, "Cotton Boll Weevil Invasion," *Independent*, Feb. 18, 1904.

cotton free zone, and no cotton was grown in the area for three years. "This district is now believed to be entirely free from the pest, demonstrating what may be accomplished where adequate control is maintained for a period of years." [24] In dealing with the problem Texas seemed particularly helpless because of constitutional limitations upon purposes for which the legislature can appropriate funds.

By 1895 the insect had reached San Antonio to the north and Wharton to the east. The drought of the next year prevented any extension of the area. In 1897 there was a slight extension of the area of insect infection. The Department sent Townsend to Mexico to look for natural parasites of the insect. He found none. The next year, the Texas legislature, having provided for a state entomologist, the Department discontinued its work and referred all inquiries to the Texas specialist. The winter of 1899–1900 furnished the lowest temperature recorded in Texas. Many expected this to finish the tropical bug, but the next spring and summer he came out later and in smaller numbers but was very prolific. By 1895 the pest had traveled two hundred miles, by 1901 he had increased his travels to five hundred and infested half the state of Texas.[25] The state of Texas had become thoroughly alarmed and had offered a $50,000 prize to the person who could propound a cure for the boll weevil. It is said that all the cranks in the state set to work, but nothing valuable came of their efforts.

Recourse was had to the Department of Agriculture in 1901, and a large sum was appropriated for experimental work leading to control. Dr. Howard, testifying

---

24 "The Cotton Situation," p. 353.

25 Dept. of Agriculture *Yearbook*, 1901, pp. 371 ff.

before the House Agricultural Committee admitted: "This is absolutely the most difficult problem in economic entomology that the whole world has ever had to handle." He went on to say that the weevil had no natural enemy, and no insecticide yet discovered was of avail. "The weevil will starve before it will eat anything but cotton." There was not a possibility of stopping its spread for its generations are capable of combined flights amounting to fifty to seventy-five miles a year.[26] Hon. George F. Burgess testified that the cold did not affect the weevil. "We have frozen some of the scoundrels in a bar of ice and kept them two days and then broken the bar of ice and put them in the sun and they thawed out and flew off." [27]

Research by government specialists and journalism had spread pictures and descriptions of the insect until it was known to farmers far and wide. Hundreds of species of weevils resemble the Mexican boll weevil. The weevil was on the public mind, and many people reported finding specimens far outside the infected area. The Department stated that the only sure way to determine that a suspected insect was a weevil was to send it to an entomologist for examination.[28] Before long, however, this grayish, or brownish bug, one-fourth of an inch long with a snout half the length of its body, was to become too well known to require expert identification. A puzzling habit of the boll weevil to play dead helped to identify it in the early days. When touched, the bug usually draws up its limbs and falls to the ground. A strange superstition arose that the weevil's long snout

[26] *Hearings before the House Committee on Agriculture Regarding the Boll Weevil*, p. 29. [27] *Ibid.*, p. 13.

[28] W. D. Hunter and B. R. Coad, *The Boll Weevil Problem*, Farmers' Bulletin 1329, pp. 5–6.

could be heated red-hot as though made of metal. The test of the truth of this must have afforded sport to many a small farmer boy.

The plant entomologists had already worked out the life history of the weevil but had gained little consolation therefrom. It was found to spend the winter in the weevil state, hibernating in stalks, weeds, underbrush, and trees at the edge of cotton fields. At the touch of warm spring days the weevils emerge to feed upon the first bolls of volunteer cotton plants. However, they can do without food from forty to ninety days after hibernation. They puncture the young squares both for feeding and for planting their eggs within. The square usually drops to the ground, and within four weeks the egg has passed from larva through pupa to weevil stage. When the bolls develop, the weevils lay their eggs in the interior just as in the squares.[29] The difficulty in poisoning the insects arises just here. They feed and lay their eggs in the inside of the boll where no insecticide, it was thought, could avail.

The factors found to limit the damage were severe winters, hot summers, and early maturing crops. Cold does kill those weevils least sheltered, as experiments showed. Eleven per cent survived in mild winters to only 2.82 per cent in severe seasons.[30] Direct, unobstructed rays of a tropic sun falling on the infected square cause death to immature weevils. In some experiments as high as 90 per cent of the weevils in larva and pupa stages have been killed.[31] Early maturing cotton simply beats

[29] *The Cotton Plant,* p. 337.

[30] Hubbard, *Cotton and the Cotton Market,* p. 39.

[31] *Ibid.,* p. 41.

the insects to it before they have had time to reproduce in sufficient numbers to ruin the crop.

The Department experts were unconvinced that the Mexican weevil did not have some native parasite, and in 1902–3 occurred a rather romantic adventure with the "Guatemala ant." O. F. Cook, sent to study the cotton culture of the Guatemala Indians, reported *Kekchi* cotton to have special characteristics that made it immune to attack by weevil. Even more interesting was his find of the *kelep*, a large, reddish brown, ant-like insect that ate weevils. Its way of dining was in this manner: With its mandibles held firmly around the belt of the weevil, the *kelep* bent its flexible body around to insert neatly a sting on the vulnerable line between head and thorax. When the poison took effect the somnolent weevil was carried to the underground nest and fed to the larvæ. After extracting the meats, the ant scrupulously and carefully packed the empty hulls away in a special underground storeroom. This ant did not eat cotton but was attracted to it by the nectar. It was in his consequent strolls over the plant that he met and took possession of the weevil.[32]

The entomologist reported that he had been found adapted for domestication and developed no bad habits in captivity. A number were captured and colonized in Texas where it was reported they bred freely. Newspaper headlines announced: "Experts Introduce Guatemala Ant to Exterminate Boll Weevil." [33] The news story was too good to be true. In its 1906 yearbook a Department specialist wrote: "The attempt to introduce the so-called

---

[32] O. F. Cook, "Cotton Culture in Guatemala," Dept. of Agriculture *Yearbook*, 1904.

[33] *Loc. cit.*

Guatemala ant to prey upon the boll weevil proved a failure."

In spite of the experimental work of the Department of Agriculture and the $50,000 prize offered by the Texas legislature, the weevil spread rapidly. Within the period from 1892 to 1922 the spread of the insect had covered the whole of the Cotton Belt with the exception of western Texas and northern North Carolina. By the end of that year 614,213 square miles of cotton-producing territory had fallen before the advancing plague of insects. Practically 87 per cent of the Cotton Belt and 96 per cent of the crop were under infestation.[34] At first the spread had been slow, but by 1900 half of Texas was infested. For the first ten years after crossing the Rio Grande the weevil's annual rate of spread was 5,640 square miles. In 1903 the weevil had reached the western tip of Louisiana, by 1906 Arkansas. The Mississippi River was crossed in 1907, and by 1910 the weevil had covered southern Mississippi and penetrated into Alabama. For the ten years from 1901 to 1911 the insect increased his annual spread to 26,880 square miles. After reaching Georgia in 1914 it spread rapidly, and in 1916 the weevil increased its flights to add 71,800 square miles of new territory. It touched South Carolina in 1917, swept across the state in two years and virtually covered North Carolina by 1923. The maps prepared by the Department of Agriculture show that the rate of spread of cotton boll weevil was much more rapid to the east than to the north or west. The droughts, as mentioned, have so far preserved the Great Plains as

[34] Hunter and Coad, *The Boll Weevil Problem,* Farmers' Bulletin 1329, p. 2.

a weevil-free cotton area. The weevil has invaded this territory but has been driven back by climatic conditions. He has shown himself wonderfully adaptable, however, and may in time adjust to the region.[35] The limiting factors in the northern advance are also climatic. The longer winters retard the ravages of the pest by causing its later emergence from hibernating quarters and by requiring a great length of time for development through all its stages. The insect thus gets a late start, and the earlier date of the first killing frost is likely to reduce its numbers.[36]

The Alluvial Belt and the Alabama Black Belt in turn fell under the onslaughts of the enemy. Insect damage ran to 9.93 per cent in 1915, 13.36 in 1916, dropped to 5.83 in 1918, and then rose to 13.20 in 1919, 19.95 in 1920, and in 1921 was over one-third of the crop, 34 per cent. It was freely predicted by British experts and secretly feared by Americans that the United States had lost her cotton monopoly and could never produce another great crop. The long staple crop in the Central Valley was practically wiped out, the Gulf Coast line was denuded of cotton, and the Sea Island crop became a thing of the past. Writing in 1923, John A. Todd, the English expert, pointed out that there was no immediate prospect of the discovery of any real cure for the weevil, suggested that it had not yet done its worst, and concluded, "The whole state of affairs . . . raises a very serious question as to the future of the American Upland crop." [37]

The seriousness with which the country as a whole regarded the menacing invasion is reflected in the heroic

---

[35] *Ibid.*, p. 4. [36] Dept. of Agriculture *Yearbook*, 1904.
[37] *The World's Cotton Crops*, pp. 105–6.

measures taken by Dr. Stubb, director of the Louisiana Experiment Station, upon finding a few weevils on the farm, at a time when the pests were not known outside Texas. The roots of the cotton plants were dug up and burned in oil with the stalks; the ground was treated with oil; and finally the Mississippi River was turned in on the farm. The fields were kept covered to a depth of two feet for ten days. Needless to say, such decisive action was rewarded by annihilation of the bugs.

It is thought that the insects found so far from their place of origin must have been placed there by some agent. Indeed, it was early foreseen that unscrupulous speculators could, by the judicious distribution of the weevil, in four or five selected places pull the price of cotton up sixty to a hundred points almost overnight. It has since become known that the Department was fearful of such a contingency and kept a strict watch.

Records of economic disaster became common history as the area of infestation moved northward or eastward each successive year. The first invasion of the weevil into new territory commonly destroyed 50 per cent or more of the crop and carried consternation to the region. The direct effects were found in the economic loss of lint and seed cotton to the growers. The consequent failures of farmers and landlords were transferred to their creditors, the banks, and supply merchants. Many of the credit institutions failed. In 1920 the Department of Agriculture estimated the average annual direct loss from the boll weevil at $300,000,000.[38] The indirect losses due to failures of merchants and bankers, the closing of cotton gins and cotton oil mills, and depre-

---

38 "The Cotton Situation," p. 350.

ciated land values cannot be estimated. They give an eloquent report, however, of the extent to which community after community had built its financial and economic structure on cotton bales.

Representative Scott Field of Texas well described before the hearing on the Boll Weevil Control Act the economic effect upon a community.

> . . . take my little town and the county in which I live upon the Brazos. I am appalled at the results that this destructive agent has produced there.
>
> The town is accustomed to ship from 20,000 to 24,000 bales of cotton. This year we will not receive 500 bales. In connection with my brother, we plant 800 acres, certainly expecting a return of 600 bales. We will not get a hundred. Men survive partial losses but when it comes so that every man looks into the face of his neighbor and is alarmed at his own condition and knows full well that he cannot discharge the obligation which the law has created; when the bank is locked like a frozen river; when no man is able to discharge the obligation that is incurred to his neighbor; when business is suspended; when land values are such that you cannot get money to make another crop and the moneyed men of the East stand in fear to invest a single dollar upon as fine real estate as there is in the state of Texas, you begin to realize the gravity of the situation.[39]

The invasion of the Black Belts by the Mexican weevil had the effect of disorganizing the whole economic and social structure. The plantation economy was disrupted and impetus was given for the Negro migrations of the war period. In certain localities the first two or three

[39] *Hearings before the House Committee on Agriculture Regarding the Boll Weevil,* pp. 36, 37.

years of infestation rendered landlords and Negro tenants almost equally helpless. "In Alabama," writes an investigator, "thousands of landlords were forced to dismiss their tenants and to close the commissaries from which came the daily rations. Some planters in Alabama and Mississippi advised their tenants to leave and even assisted them. The banks and merchants refused to extend credit when cotton was no longer to be had as security. A host of idle persons thrown suddenly on the labor market could have no other effect than to create an excess in the cities to which they flocked, make laborers easily replaceable, and consequently reduce wages. A southern paper in commenting on this situation declared, 'there is nothing for this excess population to do . . . if there is a tap that will draw off the idle population that will be a good thing for the cities at least.' "[40]

The effect of the weevil on Georgia was also to weaken the plantation and to encourage migration. Z. R. Pettit, state crops estimator for Georgia, said in his annual report for 1916: "The Negro exodus has been greatest in the territory that has been infected [with the weevil] long enough to make it difficult to grow a paying crop of cotton. The reported acute labor shortage line coincides closely with the line of third year infestation except along the southern state line."[41] Woofter also suggests that "the labor agents from the North, who were probably aware of the disorganization caused by this pest, operated more extensively in the rural district of southwest Georgia than anywhere else."[42]

---

[40] Emmett J. Scott, *Negro Migration During the War*, p. 15.

[41] Cited by Woofter, *Negro Migration*, p. 119. [42] *Loc. cit.*

Dr. Moton, President of Tuskegee, told the members of his race, "there is too much scientific knowledge, too much resourcefulness, and too much determination in the South for us to be swept off our feet by a little bug." [43] But someone, somewhere had started a song and Negroes all over the Cotton Belt were singing:

> Oh, have you heard de lates',
> De lates' all your own?
> All about de Boll Weevil
> What cosed me to leave my home?
>
> Fust time I saw de Boll Weevil
> He was sittin' on de square
> Nex' time I saw dat Weevil
> He was sittin' everywhar
> Jes a looking for a home, looking for a home! [44]

Following the shock of the first insect ravages, cotton acreage was cut down, and many regions turned through necessity to the diversified farming so earnestly preached in and out of season. The greatest immediate disadvantage to the Negroes, in fact to all tenant farmers, "was the lack of money to sustain them while corn and velvet beans were being grown." [45] The increment on live stock is not as assured as the cash returns on cotton. The means with which to begin raising live stock cannot be secured as easily as credit on cotton. But under the stimulation of war prices the South produced more "hog and hominy" in spite of its depleted man power than

---

[43] Cited in A. L. Halsey, "The Tuskegee Conference," *Social Forces,* I, 287.

[44] Dorothy Scarborough, *In the Land of Cotton,* p. 136. See also Carl Sandburg, *The American Songbag,* pp. 8–9, 252–53.

[45] Scott, *op. cit.,* p. 16.

ever before. Between 1913 and 1920 "ten southern states increased their production of corn 18 per cent, oats 50, rice 72, white potatoes 60, sweet potatoes 117, hay 150, milch cows 16, and hogs 23."[46] The people of Enterprize, Alabama, enriched by the war-time demand for the foodstuff produced by their diversified farming erected a monument with this inscription:

"In profound appreciation of the Boll Weevil and what it has done as the Herald of prosperity. This monument was erected by the citizens of Enterprize, Coffee County, Alabama."

With the close of the war came a slacking of the demand for foodstuffs followed by the economic depression and the fall in land values. In many areas the habit of diversification had been established, but in other places the Cotton Belt tended to return to the cultivation of cotton with the boll weevil added to the risks of production.

The Government experiments got under way, and specialists in fields and laboratories worked on the problem, first at Victoria, Texas, then Dallas and finally Tallulah, Louisiana. At last from the Government Weevil Laboratory, the Experiment Farm at Tallulah, Louisiana, came announcement of the long-awaited weevil cure. The patient researches of Professor B. R. Coad had discovered that the weevil, for all his boring into the depths of the boll, had a tendency to seek moisture wherever it could be found on the cotton plant. He arrived at the plan of dusting a rather virulent poison, calcium arsenate, on the plant while it was covered with dew. The dust sticks and each bath of moisture attracts

---

[46] Smith, *North America,* p. 225.

the weevil to his death. The application can be made with effect only when the plants are damp and the atmosphere is still, preferably at night.

Mr. Coad was sure he had found the answer to the cotton disaster, but the solution awaited a machine for distributing the dust. The machine method of covering large acreage was the more needed because many planters were convinced that to turn dusting by hand over to unskilled Negro tenants might mean ruin. Hand guns were worked out but proved slow, difficult to operate, and liable to breakage. They sold from $12 to $20 each, and usually lasted only one season. The first machines were failures. Many of them were made, it is said, by men who never saw a cotton field. Those with gasoline motors were unsatisfactory "because operation at night necessitated more expert labor than was available."[47] One- and two-mule machines which apply the power for the spraying device through the axle work well. The one-mule machine, operated as one would a walking cultivator, cost from $75 to $125, and was able to cover from 15 to 20 acres of cotton a night.[48]

In 1918 Coad issued his first pamphlet on the method. It aroused almost universal interest throughout the South. Calcium arsenate was difficult to make, and only one manufacturer was engaged in producing it on a commercial scale. The next year three million pounds were produced and 75,000 acres treated. The first dust contained so much water soluble arsenate as to burn the plants. When this danger was avoided the chemical sometimes failed to contain enough to injure the weevil. At Tallulah samples from farmers were analyzed. The Fed-

[47] Hubbard, *Cotton and the Cotton Market,* p. 45.

[48] *The Boll Weevil Problem,* p. 16.

eral Insecticide and Fungicide Bureau also sampled large shipments and condemned the defective lots. Machines were worked out by department designers and the models were covered by patents dedicated to the public. Blue prints of these patents were sent to all interested manufacturers.[49]

After the difficulty with calcium arsenate and machines, Coad announced two principles in the 1920 *Yearbook*. First, "Raise a cloud of dust and let it settle." Second, "The weevil can be controlled by means of calcium arsenate dust if the dust is applied at the right season at the right interval and in the right way." The difficulties exist in the fact that the control does not last long after the poison is applied and that the weevils are reduced, never exterminated. It proved useless to attempt to get weevils early in the season. The greatest difficulty was found in attempting to tell when to start and stop poisoning.[50] In 1922 Coad decided calcium arsenate was beyond the experimental stage and turned the method loose to go of its own momentum.

As weevil prevention became a standardized technique it was thought that the menace of the insect would be removed from the risks of production and become a standard item in the cost of production. The Department realized this when it advised against the use of the method unless the yield per acre is at least a third of a bale and the cost of application not more per acre than the price of a hundred pounds of seed cotton.[51]

Coad and the great Scott Plantation, Mississippi, managed by the Delta Pine and Land Company, dusted

---

[49] B. R. Coad, "Killing Boll Weevil with Poison Dust," Dept. of Agriculture *Yearbook*, 1920. [50] *Ibid.*, pp. 341–43.

[51] *The Boll Weevil Problem*, p. 15.

13,800 acres of cotton at a cost of $4.04 an acre in 1922, Later they figured the cost at $2.50 to $3.00 per acre, counting that as part of the regular crop expense.[52] Five to seven pounds of arsenate are required per acre, although weather conditions may increase the number of applications necessary. It was advised that poisoning operations be started when about 10 to 15 per cent of the squares are punctured.

Carrying his experiment further, Coad hired five army airplanes to dust 3,000 acres. It was found that the back rush of air broke the dust into clouds of fine particles, scattering it effectively. The tremendous friction electrifies the dust particles. The plants serve to ground the electric charges. Thus the particles are attracted to the plants and cling closely. In this way the fields can be dusted during the day. It is not necessary to wait for the dew. The field was dusted more cheaply than it could have been by the forty machines, forty men and eighty mules required for 3,000 acres of cotton. Congress, it is suggested, should make appropriation for further experiments. In the meantime several commercial companies have been formed for dusting cotton fields by airplane. It may prove cheapest to poison the weevil by airplane if all the farmers organize on a community basis to secure the services of the commercial company.

At the opposite end of the scale is the plan suggested for the one-horse farmer by David R. Coker of Hartsville, South Carolina. According to him, a mixture of calcium arsenate with cheap molasses can be applied by use of a mop or bottle with a shaker top. A gallon per acre is thought sufficient, and an ordinary hand can do

---

[52] Hubbard, *op. cit.*, p. 46.

four acres a day. This fits in with prevailing folkways of southern cotton culture as every member of the family except the smallest children can be used in the work.[53]

Three applications are necessary, the first when the squares begin to show. Professor Coad doubts the efficacy of the method. . . . "in many experiments," he writes, "we have yet to find where dusting with calcium arsenate is not more profitable and more effective than the use of poisoned sweets."[54] The method will likely survive, however, simply because in the southern cotton field human labor is cheaper than machinery.

A method has also been worked out in Florida at the Experiment Station by G. D. Smith. It has proved efficient at a cost of only $1.57 per acre. It depends upon knowledge of completion of spring emergence of weevil and consists in removing and destroying all the squares on the plant at a time when about two per plant are present. If followed by dusting of the terminal buds many of the pests are destroyed.[55]

In February, 1923, the Southern Agricultural Workers at the Memphis convention indorsed the calcium arsenate and Florida methods and called for new tests on the molasses-arsenate treatment. At the same time the Department issued warnings against patent anti-boll weevil medicines.[56]

The risks of the weather are as great for cotton as any other major crop. Because of the southern climate, droughts and excessive rains are very real dangers to special localities.

---

[53] Alfred G. Smith, "Winning the Weevil War," *Country Gentleman,* May 31, 1924.

[54] Cited by W. H. Hubbard, *op. cit.,* p. 46.

[55] *The Boll Weevil Problem,* p. 30. [56] *Ibid.,* p. 1.

De drouf hit parched our crop at fust
En de rain done drown it now.

. . . .

De cawn jis want some 'scuse to quit,
En cotton's a regular chile
But de sun kin scawch and de rain can spit,
But de crab grass wears a smile.

In the Mexican boll weevil cotton possesses a scourge greater than that attacking almost any other crop. Its total destruction of cotton has been without a doubt immense. The opinion, however, is held by some agricultural economists that the boll weevil has been the economic salvation of the industry. When it has brought better prices to the crop by keeping down overproduction and by forcing many farmers to diversify, it has accomplished that which generations of preaching and teaching had failed to do. It has even been said that as long as cotton acreage remains uncontrolled, seriously to wage war against the boll weevil is economic treason. Regardless of the fallacy in raising crops to feed bugs, the argument would be sounder had the weevil loss fallen with equal severity upon all the growers. It is commonplace to point out that high prices for cotton in one section have been attained at the expense of farmers who have had their crop almost totally wiped out in other sections. The reduction has not been evenly distributed.

The future of the weevil in the American cotton fields is difficult to predict. In the last few years there has existed the tendency to depend more upon cultivation and "forcing" methods and less upon doctoring in attempts to beat the weevil. The weather is easily the most important and the most variable factor. Conditions of exceptionally dry weather and hot sunshine during most of the growing season produced bumper crops for 1925 and 1926 with higher yields per acre than for any recent period. It began to be hoped that the weevil had been conquered. But the less exceptional weather conditions of 1927 produced a low yield and apparently found the weevil back at his old tricks. The suggestion is obvious. The weather has more effect in controlling the boll weevil than our boasted science. It seems more likely than ever that the weevil is here to stay. There exists no way at present to attack him except through laborious methods of forcing cultivation or applications of calcium arsenate. His continued presence and his dependence upon the vagaries of climate will serve to accentuate the effects of rains and droughts. Thus the weevil will decrease production at unforeseen periods and will leave it practically unmolested at others. To the extent the weevil does this he will emphasize price fluctuations. Cotton bids fair to continue to deserve the epithet of dynamite. Our next chapter, accordingly, will discuss the hazards of the market to the cotton producer.

CHAPTER V

THE RISKS OF THE COTTON MARKET

SPECULATIVE NATURE OF COTTON RAISING

We have just seen that on the producer of cotton falls the burden of two great risks of nature: the weather and the weevil. To the vicissitudes of production must be added another hazard just as great—the risks of the market. The great variations in production suggest wide variation in prices. What economists call that "imperfectly adjusting and readjusting equation of supply and demand in a money economy" is nowhere better exemplified than in the cotton market of the world. The changes in production due to wars, to depredations of insects, and to the vagaries of the American climate have made cotton the humpty dumpty of commodity crops. The study of cotton supplies, cotton surpluses, cotton consumption, and cotton prices is the lifetime work of experts engaged in the classification and distribution of stocks of raw cotton to spinners. Consequently, most of the studies devoted to fluctuations in cotton have been made from the standpoint of the consumer and manufacturer. It is the purpose of this chapter to address itself to the interrelations of average producer and price fluctuations.

The raising of cotton is essentially a speculative industry. "As I have no disposition to gamble or invest in lotteries, I do not grow cotton," wrote an Arkansas planter in the 1860's.[1] The feeling was more forcibly expressed by a North Carolina pamphleteer writing in the picturesque phraseology of 1883:

> There is another view of this subject, we now call to the attention of the farmers, thousands of whom are earnest and devout Christians.
>
> The man who will jeopardize the happiness of his wife and little ones and deprive them of bread, by a cast of the die, or the turn of a card at the faro table they deem lost to all feeling of morality and swiftly bound for hell.
>
> Yet, year after year, they hazard the happiness and livelihood of their fond wives and children by giving mortgages upon their homes and liens upon their crops, "trusting to chances," that they will make a good cotton crop and be able to pay them off, that "other countries will fail in the production of cotton this year, and ours will, consequently, be higher," that labor will be cheaper than last year, that there will be not too much nor too little rain, that there will be no hail storms, and a hundred contingencies, the happening or not happening of which would be sufficient to take the bread out of their families' mouths and deprive them of a home.
>
> [This] "trusting to chance" . . . is a higher species of gambling than that of the man who bets $10 that a 48 will be thrown or the ace turned; in this respect the two differ. Do they differ otherwise? Let each man answer for himself.[2]

Within the memory of many southern farmers the price of a standard five-hundred-pound bale of cotton has varied from $35 to $200. Taken year after year, it is doubtful, as Hammond says, whether cotton yields as large returns as would result from a safer, more diver-

---

[1] Cited in Hammond, *The Cotton Industry*, p. 87.

[2] W. R. Henry, *Cotton and the Commission Merchants*, p. 21. Pamphlet.

sified system of farming.[3] As is so often admitted, cotton proves the occasion of serious loss. But high prices when they come remove the gloom, and Cotton the King is again in high favor with his chief subject, the planter.

In one way the risks of the market fall with peculiar directness on cotton. Of the important crops produced in the United States, cotton is peculiar in that all of the crop must be sold by those who produce it. The only use of cotton to the cotton farmer is its use in exchange, its only value is its market value. Cotton thus is governed completely by a money economy. In fact, the lower the price of cotton the greater likelihood that the farmer will have to sell immediately all he has produced to satisfy his creditors as speedily as possible. On the other hand, it is estimated that only 17 per cent of the great United States corn crops ever leave the farm where they are produced. The live-stock farmer may sell corn and oats, or he may sell hogs and cattle. Corn, pork, and beef fill the family larder regardless of the price. The all-cotton farmer possesses nothing that he can use; he must sell it all. The risks of the market do not fall with such crushing force upon the grower of corn. Correspondingly, the need for diversification is greater with the cotton grower.

On the other hand, the cotton industry of America practically possesses a natural monopoly such as obtains in no other product of equal importance. American production, which averages around 60 per cent of world production, practically determines the international price of cotton. American Upland Middling, which is expected yearly to make up 90 per cent of the United States crop,

---

[3] *Op. cit.*, p. 87.

furnishes "the 'basic price' which controls the prices of all other grades of American Upland cotton and also, more or less, of all other kinds of cotton."[4]

In April, 1924, a gloomy period for cotton manufacturers because of the short crops of the three years preceding, John A. Todd, the English authority, was quoted in a leading American farm journal as seeing no chance of increasing supplies outside the United States over a million bales a year. "One week of favorable or unfavorable weather at a critical time in America means more to the world's immediate cotton supply than all the efforts being made to increase production elsewhere," he said.[5] In *The Cotton World* published in 1927 he made the statement, ". . . that the fluctuations of the American crop from year to year are so great that they may easily counteract the total effect of any changes that take place in all the other crops put together."[6]

No other group of American agricultural producers have an economic position of such strategic value. The United States may have a short wheat crop, for instance, with very little influence on the price. An example will serve. With a United States crop of 374 million bushels in 1893, wheat sold for 30 cents a bushel less than in 1891, with a United States crop of 600 million. To what extent a short American cotton crop may influence world prices was shown in the effects of the localized American Civil War. In 1863 the lowest spot price of middling was quoted at 70 cents and the highest at $1.60.[7] The export price is given at 55 cents, but there was no cotton

[4] Todd, *The World's Cotton Crops*, p. 1.

[5] E. H. Taylor, "Cotton Is Sitting Pretty," *Country Gentleman*, April 19, 1924, p. 9.

[6] P. 6.

[7] "The Cotton Situation," Fig. 43, p. 389.

for export owing to the blockade.[8] R. G. Engberg in his study found the size of the American crop to have a high negative correlation with world cotton prices. "If we take percentage by which the New York spot price in December changes from that of the preceding year for each year from 1881 to 1913 and correlate these figures with the amount of cotton ginned in corresponding years, a correlation of —.778 is obtained."[9] He concluded that this shows the volume of American production is the major price factor.[10]

The impression should not be given, however, that this potential monopoly has been of any advantage to the producers. Such fluctuations in price as have occurred have been due mainly to natural causes affecting production. Cotton growing is a most disorganized industry, and cotton growers are extremely individualistic. There are no economic and social policies developed to strengthen the grower's position. Many southern writers and orators have noted this fundamental contrast between the potential and the actual situation of the cotton grower. The following quotation from an address of W. B. Thompson, President of the Southern Cotton Association at its convention in Dallas, Texas, in 1908, is a type of statement that has been uttered so often in the South that it has no meaning:

> The cotton growers of the South are the greatest wealth producers in the world. The cotton growing industry is potential to make the South the richest country in the world. In years gone by, because of helplessness awhile, and because of the habit of submission, the South has realized the

---

[8] *Cotton Atlas,* Fig. 68, p. 20.

[9] *Industrial Prosperity and the Farmer,* p. 152. [10] *Loc. cit.*

> minimum benefit from its great opportunity, and like a blind and listless Dives [*sic*] has accepted the crumbs while the body of the feast has been swept away and fed to others. Let the cotton planter realize the possibilities of his advantages and it will follow that every other business interest of the South and every acre of land and every share of stock in business will increase in value.[11]

Every bad year is the occasion, as we shall see, of social unrest and innumerable panaceas for relief. All these are forgotten upon the next good price year, and so the cycle goes. Campaigns for restriction of cotton acreage follow upon the wake of every disaster. Agreements to cut acreage 10, 15, or 20 per cent are cheerfully made, and in many instances as cheerfully broken. One almost comes to feel that the speculative element in cotton possesses a fatal fascination for the farmer. One good year offers balm for several bad ones. Many intelligent farmers can be found who prefer to risk their crops on outguessing their neighbors in planting and producing cotton rather than join with them in coöperative determination of cotton acreage. There are other important factors in this connection. The Cotton Belt is too big to be successfully organized. There exist, as will be shown, differential costs of production in favor of the farmers of the new western areas. But the idea remains that one never knows what cotton will bring and one cannot trust one's neighbor. And after all there is a chance to beat the game, for "What do you 'spose cotton'll bring next fall?"

The interaction of demand and supply as a causative factor in determining cotton prices is a problem de-

11 Address at Annual Convention, February 24, 1905, pp. 1–2.

manding elaborate statistical analysis by the economic technician. It has been carefully presented by Professor H. L. Moore in a monograph on *Forecasting the Yield and Price of Cotton.*[12] It is sufficient to say that demand as conceived by the economist is a very elastic thing, absorbing large crops offered at low prices in a surprisingly short time and measurably reducing the rate of consumption of small crops offered at high prices. New uses for cotton, as in automobile tires, are likely to be discovered at any time. Higher standards of living and new social customs among remote peoples are an unpredictable though no less important factor. The increasing use of silk and the advent of new artificial fabrics like rayon cause a lessening in demand. Owing to the vanishing forests, rayon will have to be made from short lint cotton. In the long run the manufacture of rayon may come to be regarded as the substitution of an expensive technical process for nature's chemistry of soil and sunlight. But increased uses of cotton throughout the world await the raising of the standards of living of many peoples. As Honorable Adam M. Byrd of Mississippi told his fellow-members of the House, the cotton farmer's way to prosperity depends on his inalienable right "to displace the fig leaf in all the ends of the world by the cotton shirt."[13] It was Ambassador Wu Ting Fang of China who charmingly and hopefully admonished his hosts, the Southern Textile Manufacturers, that if his countrymen should decide to add another inch to the tails of their shirts it would require the whole cotton crop of the South.

---

[12] Published in 1917.

[13] Address in U. S. House of Representatives, Feb. 24, 1905.

THE TREND OF PRICES

Cotton prices are the barometer of this ever changing relation between supply and demand. From the beginning of cotton culture in America they have shown a varied range. Prices were rising when Whitney invented the cotton gin in 1793, and by 1901 the consequent increased production had taken effect, causing a drop from 44 to 22 cents per pound export prices.[14] Prices continued to fall for the next decade owing to the embargoes preceding the War of 1812 and the spread of cotton culture through all the South Atlantic states. Cotton reached its lowest at 11 cents in 1811, but the War of 1812, by hindering the importation of cotton goods, helped to establish the cotton manufacturing industry in America. This new industry being protected by tariff, the price of cotton rose steadily until it reached 34 cents in 1817. The high prices had doubled production during 1815–18 in a period of rapid westward expansion to Alabama, Mississippi, and Louisiana. Under the pressure of increasing production prices fell steadily (excepting the short crop year of 1824) until 1830 when they reached the low level of 9 cents. The market demands of both foreign and domestic manufactures were keeping pace, however, and prices rose steadily until 1838, reaching levels between 17 and 14 cents. In this period production had doubled again and the United States was growing eight times as much cotton as in 1815. High prices, removal of Indians, and consequent opening to settlers of new lands in Georgia, Alabama, and Mississippi, produced speculation, overexpansion, and overproduction which continued to depress prices until 1844. Although during

[14] *Cotton Atlas,* p. 20.

this period cotton sold as low as 6 cents in 1844, the production increased 50 per cent. From 1844 to 1859 was a period of rising prices. The South was prosperous in the decade from 1849 to 1859, and cotton averaged over 10 cents, reaching 12 cents in 1850. Portions of Texas and Arkansas were opened up, and as U. B. Phillips has shown, the railroads from the coast to the interior further increased production in inland North Carolina, South Carolina, Georgia, and Alabama.[15]

The Civil War marks a peak in the price of cotton. The blockade and temporary stoppage of production ran the price up to various levels reported at $1.80, highest spot price, 70 cents, lowest spot, and 55 cents export price. The crop of 1861 had been the second largest thus far produced, and much of this became available after the war at high prices. In 1865 over 2,000,000 bales were marketed at a price of 30 cents. Due to difficulties of political and industrial readjustments with the human factors, it was not until 1877 that pre-war production was reached, although over four million bales were produced both in 1870 and 1875. Prices fell rapidly until they reached 10 cents in 1878. For the twenty years from 1879–99 production doubled, and prices continued to fall. Prices from 1891 to 1902 never saw 10 cents, and in 1898 fell to the lowest point in history, 5.6 cents. The view, sometimes expressed, that if conditions made the cultivation of cotton wholly and prospectively unprofitable, it would take a whole generation to wean southern farmers from its culture, received adequate proof in the 1890's. A slacking in the rate of spread of the cotton area resulted in rising prices, which

[15] *A History of Transportation in the Eastern Cotton Belt to 1860*, chap. I.

hung around 10, 12, and 14 cents until the 1914 crop. In that year the largest crop in history and the disorganization in the cotton markets attendant upon the outbreak of the World War broke the market to 6.8 cents. In 1914 "cotton brought tears" to the whole South. Increasing war demands again raised the prices, which reached as high as 35 cents in 1919, the highest since the Civil War. In 1920 the price depression carried cotton with a crash down to 16 cents. Resulting reduction of acreage and boll weevil damage began a rise which carried it back to 23 cents in 1924. Since then the largest acreages in history reduced the price until it reached 10 cents in 1926, lowest since 1915. The pendulum swung again, and the 1927 crop is estimated to have brought the farmer from 18 to 20 cents.

The foregoing history of cotton prices indicates that cotton growing is a more or less haphazard game of see-saw between production on one hand and demand as represented by prices on the other. A generalized type description of their interaction would run somewhat as follows: Good prices the first year stimulate larger plantings the next year. The second year the increased acreage sends prices down. These low prices discourage overproduction the third year. Comparatively low acreages plus increasing demands, and new uses for cotton cause good prices and the cycle is complete. "The cotton cycle," writes R. G. Engberg, "is normally two years in length. A large crop is usually accompanied by a relatively low price. This tends to reduce the acreage so much the next year that unless there is an abnormally large yield the total crop also is materially reduced. The price therefore rises again and the cycle is complete."[16] Add to this cycle the complex, unpredictable factors of weather and weevil and one understands why the cotton farmer is said to hold a ticket in the world's chief lottery.

Bradford B. Smith writing in the *Journal of the American Statistical Society* states the economic theory involved:

> The producers of agricultural commodities will, when taken in the mass, plant that product which they believe will yield them the largest net returns or profits. One of the chief factors influencing the net returns is the selling price of the selected crop relative to the possible crops. Thus if the producer is of the opinion that there will be a relatively high price for his cotton crop, that is relatively higher than corn, for example, he will desire to plant cotton rather than other crops; and the greater the relative price the more [cotton] he would wish to have planted.[17]

Hubbard writing from the viewpoint of the cotton trade comments:

> There is an agitation each year based on the one crop theory, but the trade pay little or no attention to editorials and speeches on the subject, for they know from past experience that it is the price of cotton, the price of supplies, the supply of labor, and other economic facts, which in the end determine the acreage. The southern farmer and merchant know this, also, but they enjoy the annual discussion which they feel develops the facts.[18]

This psychological theory of the cotton producer as "economic man" choosing and rejecting among elements

---

16 *Industrial Prosperity and the Farmer,* p. 57.

17 "Forecasting the Acreage of Cotton," March, 1925, p. 32.

18 *Op. cit.*, p. 20.

of production may be true in the main. It neglects, however, important social and racial factors and lumps all the human factors in cotton production together in one generalization. The cotton cropper, the permanent tenant, the farm owner-operator, and the landlord vary essentially in their reactions toward cotton prices as F. W. Gist points out.[19] The cropper, owning nothing but his labor, is bound to his landlord, and both are wedded to cotton. "The landlord is not interested in making money in cotton, but in getting his rents in cotton. Cotton means cash; the cropper can neither steal it nor feed to stock. The cropper is interested simply in paying his rent and his store bill so that he can eat again next year. It is estimated that this group, Negro and white, produce 25 per cent of the cotton grown. This group is never interested in economic production, or in preventing overproduction except when the price of cotton is too low to cover rent or pay store bills."[20] A step above the cropper, the share renter is estimated to grow about 40 per cent of the cotton crop. He may vary his acreage more than does the cropper, but both southern croppers and share renters, aided and abetted by their landlords and supply merchants, tend to grow cotton without regard to the price. They do not conform to the theory of the "economic man."

The farm owner-operators in the cotton states are estimated to grow about 35 per cent of the cotton crop. They grow food and possess a cash income beside that from cotton. It is they who restrict acreage when prices are low and chase high prices with increased acreage.

---

[19] "Practical Crop Diversification in the Cotton States," *Commerce and Finance*, Jan. 12, 1927, p. 115.

[20] *Loc. cit.*

If owners are forced to the need of credit they can secure it by giving a crop lien on cotton. The variable human factor then in the production of cotton, the one economic man, is this owner-operator. He is able to speculate with the cotton market by varying his acreage, and it is his reductions that help to vary production. Working on the problem of the relation between the prices of agricultural commodities and the acreage devoted to that crop the next year, R. G. Engberg of the Institute of Economics found that the highest relation existed for cotton. A correlation of +.62 with a probable error of .08 was found to obtain between the price of cotton and the acreage planted to the crop the next year. The only other significant correlation found, +.42, was for flax.[21]

As a matter of fact, the reductions of acreage do not fully account for the price fluctuations. When the cotton farmer reduces acreage, nature decreases the production in a larger ratio and the price of cotton is correspondingly increased in greater proportion. That this tendency exists is shown in the analysis of the following table compiled and commented on by Carl Geller.[22]

A glance at the table brings out the following facts: Since 1890 there have been nine low-priced bumper crops followed by nine attempts to cut acreage. The highest reduction in acreage was 17 per cent, the lowest 3, and the average 11. The corresponding reduction in size of crop ranged from 13 to as high as 40 per cent, with the average at 25.2. The resulting increases in price ranged from 8 per cent (owing to an excessively large

---

[21] *Industrial Prosperity and the Farmer,* p. 34.

[22] "Fewer Acres and More Dollars," *Commerce and Finance,* Jan. 12, 1927, p. 123.

surplus in 1921), to 64 per cent, with the average at 33 per cent.

TABLE IV

Year	Percentages of Reduction of Acreages	Percentages of Reduction in Size of Crop	Percentage of Advance in Average Yearly Price to Farmers
1892–93	17	25	15
1895–96	14	29	39
1899–1900	3	17	35
1905–06	15	20	25
1909–10	3	24	55
1912–13	6	13	20
1915–16	14	30	51
1921–22	14	40	8
1926–27	13.6	29.8	64

For these eight years an average reduction of 11 per cent in acreage has meant an average of 25.2 per cent reduction in the crops, resulting in an average 33 per cent rise in cotton prices. The effect of the reduction of acreage has, by the time it reaches prices, been greatly accelerated. Thus the original variation of 11 per cent has been tripled in the final outcome. Engberg in his calculations reaches the same conclusions, finding the yield more variable than the acreage. "Of the two factors, yield and acreage, which make up total production, yield varies the more widely. The coefficient of variability of the percentage changes of yield is 14.60 per cent, and that of acreage 9.03 per cent." [23]

What is the explanation of this tendency? The production of a bumper crop is due to large acreage, plus an equally important factor—exceptionally fine weather. There are likely no observable cycles of weather, and

[23] *Op. cit.*, p. 159.

there may be no laws of chance applicable, but an exceptionally fine season is rarely followed by another so favorable. If the weather holds the weevil in check one year, it is not likely to be so kind next year. Decreased acreage thus meets less favorable weather conditions or increased weevil damage, and the total effect is passed on to raise the price all out of proportion to the original reduction in acreage. Such a tabulation serves to show only too clearly the speculative nature of cotton growing.

It can be said in no facetious spirit that all unwittingly the program of reduction of the cotton acreage has been too successful. It may be asked what has been the effect of the rises in price after reducing acreage. It has simply made impossible any long-time scheme of stabilizing cotton production in the South. No general plan of diversification, it seems, is able to stand unmoved before rising cotton prices. A less sharp cutting down of production, followed by slowly rising prices, would be more conducive to a correct apportioning of the southern farm to cotton.

The contrast between society's interest in adequate supplies and the producer's interest in an adequate return on production is nowhere better shown in our competitive money economy than in the cotton market. The speculative element plus the "transvaluation of all values into money values" has produced that paradox in economics—the larger total price of a smaller crop.

Seven times since 1890 a decrease in production has occurred and five times resulted in an increase in the total money value of the cotton crop. The following table [24] is arranged to show the paradox of value:

---

[24] Geller, *op. cit.*, pp. 123, 125.

TABLE V

YEAR	CROP MILLION BALES	VALUE MILLION DOLLARS	YEAR	CROP MILLION BALES	VALUE MILLION DOLLARS
1894–95.....	10.0	$295.0	1895–96....	7.1	$291.1
1898–99.....	11.4	282.8	1899–1900..	9.5	363.8
1904–05.....	13.4	561.1	1905–06....	10.6	556.8
1908–09.....	13.2	588.8	1909–10....	10.0	688.4
1911–12.....	15.7	750.0	1912–13....	13.7	786.8
1914–15.....	16.1	591.1	1915–16....	11.2	627.9
1926–27.....	17.9	982.7	1927–28....	12.7	1,253.5
	97.7	$4,051.5		74.8	$4,568.3

An absolute decrease of 22.9 millions bales has meant an absolute increase of $406,800,000 in the total value of seven crops. A decrease in production of 23.4 per cent is thus shown to bring a total increase in the money value of the crop of 12.9 per cent.[25]

To the southern agricultural interests this paradox of value appears as veritable ruin because of plenty. It rankles in the breast of the cotton grower as an unmerited punishment for the answer of his industry to the world's demand for more cotton. This view is given typical expression in a statement by the general manager of the American Cotton Growers' Exchange:

For four years there has been a practical famine of cotton and the growers have been urged to increase production by expensive methods of insect control, crop fertilization, and intensive cultivation, with the result that they have succeeded in meeting the world's need with a small surplus or reserve in case of future insufficient cotton yield. The cotton trade

[25] The years 1892–93 are omitted because they are exceptional years—the first because of the great Lancashire strike; the second because of the world-wide depression.

and society apparently have answered the successful effort of the farmer with less dollars for a 15,500,000 to 16,000,000 bale crop than he was paid for approximately 2,000,000 bale smaller crop last year. In the face of this discouragement, with the resulting prices below the cost of production even for the better grades and with terrific losses on the lower grades which cost practically as much to produce, there is nothing for the farmer to do but reduce his acreage and produce only such quantity of cotton as the world appears to desire. Failing in this method of securing profitable prices the cotton farmers like wheat farmers will be compelled to turn to governmental price controlling expediences, even though in the past agricultural leaders of the South have opposed such methods of stabilizing prices at fair levels. It appears that the cotton trade does not desire reasonable reserves as they penalize farmers when they exist.[26]

Although they are often given, neither fluctuations in price per pound nor total values of the crop accurately relate the average producer of cotton to the hazards of the market. An index of the risk of the producer must combine two figures; viz., the yield of the crop, large or small, multiplied by the average price. This figure may be secured by multiplying the average yield of pounds of lint per cotton acre by the average farm price per pound. The following table from 1909 to 1927 is based on the Department of Agriculture figures.[27] From 1889 to 1909 the statistics are compiled from Lea's *Cotton Book*.[28]

---

[26] C. O. Moser, *Commerce and Finance*, Jan. 13, 1926, p. 121.
[27] Dept. of Agriculture *Yearbook*, 1925–27, p. 952.
[28] Pp. 24, 26.

TABLE VI

AVERAGE RETURN TO PRODUCER PER ACRE OF COTTON IN UNITED STATES—1878–1927

YEAR	AVERAGE YIELD PER ACRE IN POUNDS OF LINT	AVERAGE FARM PRICE CENTS PER POUND	AVERAGE RETURN TO PRODUCER PER ACRE OF COTTON
1927	152.3	19.6	$31.21
1926	182.6	10.2	20.87
1925	167.2	18.2	31.79
1924	157.6	22.6	37.26
1923	130.6	31.0	42.34
1922	141.3	23.8	35.14
1921	124.5	16.2	21.11
1920	178.4	13.9	20.02
1919	161.3	35.6	60.62
1918	159.6	27.6	46.20
1917	159.7	27.7	46.28
1916	156.6	19.6	32.08
1915	170.3	11.3	20.10
1914	209.2	6.8	14.91
1913	182.0	12.2	23.26
1912	190.9	11.9	23.83
1911	207.7	8.8	19.08
1910	170.7	14.1	25.32
1909	154.3	13.9*	22.55
1908	206	9.45	19.27
1907	177	11.42	20.21
1906	211	9.56	20.17
1905	209	11.07	23.04
1904	216	8.98	20.51
1903	172	12.15	20.90
1902	190	8.82	16.76
1901	192	8.06	15.48
1900	193	9.33	18.01
1899	185	7.65	14.99
1898	221	4.88	10.78
1897	222	5.64	12.42
1896	172	7.32	12.71
1895	169	8.18	13.93
1894	202	5.92	11.96
1893	183	7.50	13.72
1892	176	8.40	14.78
1891	206	7.30	14.48
1890	177	8.60	15.22
1889	173	11.07	19.15
1888	173	10.65	18.42

*From 1909 to 1878 the figures show the average price per pound received by producers December 1st.

TABLE VI (*Continued*)

Year	Average Yield per Acre in Pounds of Lint	Average Farm Price Cents per Pound	Average Return to Producer per Acre of Cotton
1887	177	10.03	17.75
1886	164	10.21	16.74
1885	166	9.28	15.40
1884	156	10.45	16.30
1883	157	10.88	17.08
1882	195	11.88	23.17
1881	159	11.56	18.38
1880	196	12.03	23.58
1879	208	11.51	23.94
1878	185	10.84	20.05

The table reveals some interesting facts as to the income of cotton farmers. Within the last fifty years the income per cotton acre has ranged from $10.78 in 1898 to $60.62 in 1919. In the decade of 1890 the returns per acre reached as high as $15.00 but once. Owing to war conditions, boll weevil, and new uses for cotton, the decade 1916 to 1926 has offered the highest returns per acre since the Civil War decade, the lowest $20.02 being in 1920, year of depression and the mean being around $32.28. During this period the cost of production has also been at its highest. Accepting the estimate commonly made that the average cropper, share tenant, or small owner, cultivating virtually no other crop and working only himself and family, can tend approximately twenty acres of cotton, we can see that within a generation his money income may have varied from $215.60 to $1,212.40 a year. To know the effects on the average producer we should study the varying costs of production and his standards of living. In this chapter, however, we are concerned with the cyclical fluctuations, movements of prices, and the speculative nature of cotton growing.

The effect of fluctuations of cotton prices on the southern community is complex and far-reaching. Within themselves these price movements may be said to constitute practically another business cycle for the South. The business cycle is one thing; the cotton cycle another. Engberg attempted to investigate the price influence of the domestic business cycles on cotton. Using the volume of pig iron production as an index of general business conditions, he correlated these figures with the price of cotton nine months afterward. He found a correlation of +.489.[29] Below .50 is not regarded as a high correlation. It will be remembered that the correlation of the size of the domestic crop to price was found to be −.778, much more significant. He sums up: "It was found that out of 19 farm products, including butter and cheese, only three—hogs, sheep and cotton—had fluctuations at all typical of the business cycle." [30]

The coefficients indicate that only in the case of cotton does the apparent influence of the business cycle approach in magnitude the influence of the volume of production. Often moving in harmony with the national cycles, cotton prices have frequently operated to give the South a period of depression in the midst of the nation's prosperity. It might be pointed out that while a nation-wide depression would be reflected in reduced takings of cotton mills, a corresponding period of prosperity would not take up the slack in a bumper cotton crop. A like phenomenon is found in the Western Wheat Belt. It is the penalty visited upon specialized regions. A good example

[29] *Industrial Prosperity and the Farmer*, p. 154.

[30] *Ibid.*, p. 86.

of commodity depression during industrial prosperity is the stupendous crop of 1926, 17,977,000 bales, possibly the largest ever grown, which sold for 10.2 cents. The result was to plunge the Cotton Belt into a period of depression in spite of the fact that the business cycle for the nation registered prosperity. "The cotton crop of 1926," the President of the Georgia State Agricultural College estimated, "will bring the farmers who produce it $1,000,000,000 less than they needed in order to break even or slightly better." [31] This loss "represents $1,000,000,000 of frozen loans and unpaid bills which cannot be liquidated."

COMPETING COTTON AREAS

Not only is production fluctuating, but cotton acreage tends to increase in spite of falling prices. This additional hazard to the cotton producers results from the competition between different cotton areas. It can be shown that the Western Belt can produce cotton at a profit while eastern farmers are growing the crop at a loss. From 1919 to 1924 in a period of falling markets in commodity crops, the wheat area decreased 22.2 millions of acres; at the same time cotton increased 10.2 millions of acres. It is likely that there exists a surplus of 10 million acres of cotton with no way of reducing it except by abandoning farm land. At a time when the cotton grower in the Gulf and Eastern Belts is likely selling cotton under cost of production, acreage is being increased in newly developed regions of western Texas, western Oklahoma, and eastern New Mexico.

---

[31] Andrew M. Soule, "Problem of the Cotton-Growing Industry," *Commerce and Finance*, Jan. 12, 1927, p. 109.

"Abandonment of cotton farms," writes W. J. Spillman, "is going on rather rapidly along the eastern Gulf Coast and the South Atlantic Coast. . . . At the same time, the acreage of cotton is being extended northward and particularly westward on the plains of Texas and New Mexico."[32] The opening up of a new area in which the possibilities of cotton culture had not been anticipated has been accompanied by cheaper costs of production for Texas, a rapidly increasing acreage, and a fluctuating yield that tend to play havoc with the market. The competition between the Eastern and Western Belts has thus added more hazards to the industry. "With the cotton area contracting on one side and expanding on two other sides, the stabilization of cotton acreage presents great difficulties."[33]

The Director of the South Carolina Experiment Station in 1925 called attention to the losing competition with the western farmer. "When the yield [of cotton] goes as low as 150 pounds per acre, which is the average for the entire state for the last five years, and cotton sells for 18 to 20 cents per pound or less, the crop is produced at a loss. The eastern part of the Cotton Belt is now meeting keen competition from the states farther west where on account of weather conditions, more fertile soil, and low boll weevil damage the cost of production is much less than in this section."[34] A large part of the impact of this competition has resulted from the opening

---

[32] *Balancing the Farm Output*, p. 49.

[33] *Ibid.*, p. 50.

[34] *Annual Report*, 1925. Cited by D. W. Watkins, *An Investigation of Cotton Growing in Texas Showing Influence on Cotton Production in South Carolina*, p. 2.

up to cotton farming of a hitherto undeveloped region in the Great Plains.

From 1919 to 1926 Texas increased her cotton acreage from over ten million to over eighteen million and her production from three million to almost six million bales. A great part of this increase occurred in a region which had been considered beyond the possibility of cotton culture, the Staked Plains. In 1918 the Department experts had written in the *Cotton Atlas*: "The expansion of the Cotton Belt has approximately reached its climatic limits, and future increases in acreage must be mostly by a more complete utilization of the land within these limits." [35] This prophecy was to receive a striking contradiction. Located near the south end of the Great American Desert, the Texas Staked Plains first furnished winter range for the buffalo and after 1865 became the scene of western cattle ranches.[36] Under the impression that such land would never be available for crops, stock raisers acquired ranches comprising thousands of acres by what practically amounted to violation of Texas land laws. By 1885 it was shown that crops could be grown in this area, and the small farmer began to push his way into the midst of the ranching economy. Between 1919 and 1924 a million acres were added to cultivation.[37] The decline in the value of live stock and the rising price of farm land had practically completed the removal of cattle raising from the region by 1925.[38]

The soil has been described as "a heterogeneous mixture of outwash material segregated to some extent into

---

[35] P. 23.

[36] E. O. Wooten, "Cotton in Texas Plain Area," Dept. of Agriculture *Yearbook,* 1926, pp. 271–74.

[37] *Ibid.,* p. 271.

[38] *Ibid.,* p. 272.

coarser and finer portions by river and ocean action and at one time was the bottom of a shallow ocean. The material was largely consolidated into sandstone, shale, and limestone, was afterwards elevated above sea level and has since been covered to a certain extent by outwash from material and by wind-blown deposits of sand and silt."[39] "The land above cap rock is generally level and gently rolling and the tillable soils are mostly light and easily worked when properly moist; hence, cultivation of large fields with large equipment is easy. These new soils are now fertile; hence, the application of fertilizers is not necessary at present and it is possible that on some of them it never will be."[40]

The introduction of agriculture into this region waited upon a substitute for corn, which was unable to stand the dry hot weather. This was found in the cultivation of kafir corn, milo, and sudan. The introduction of cotton in this area is said to have been a surprise to everyone except those concerned in its cultivation. By 1924 an acreage of 1,269,600 in Texas and 22,700 in New Mexico were planted. "When once new land [in the Staked Plains] is plowed and put in cultivated crops, it rarely is used again for grazing as it does not become reset in the native grasses satisfactorily for many years."[41] Thus the land is kept in cotton or a competing crop.

In the Staked Plains as a whole, it is estimated that over fifteen millions of acres in Texas and four and a half in New Mexico may be used for crops and pastures. Under favorable market conditions it is likely that from

---

[39] Milton Whitney, *Soils and Civilization,* p. 78.

[40] Wooten, *op. cit.,* p. 272. [41] Watkins, *op. cit.,* p. 5.

18 to 20 per cent of the region would be devoted to cotton, giving about 3,200,000 new acres.

The advantages of cotton culture in this area rest upon the fact that high altitude and dry climate have so far kept the boll weevil in check. Thus the necessity for continuous cultivation through the season is not felt. This advantage is further aided by the absence of troublesome weeds. The level fields allow the riding machine type of cultivation to be used to the utmost. A farmer riding a two-row lister with a six-mule team and a set of fenders, knives, disks, and points can plow and plant the land at one operation. Tractors and plowing teams of four, six, and eight horses are used. Under climatic influences the plants tend to ripen at almost the same time. Thus there has grown up a mechanized method of picking cotton called "sledding." These sleds are run across the level fields, plucking the fruited fiber and unopened bolls. By means of specially perfected machinery the boll cotton is opened, and the scraps of foliage extracted during the process of ginning. Accordingly, so much less labor of men and mules is required that one farmer may cultivate as high as one hundred acres in cotton. Cotton picked with the cotton burr to save time is known as "snaps," and frost bitten bolls harvested whole are called "bollies." The gins return remarkably clean bales from snaps, bollies, and sledded cotton, but the grade is not as good as that of the cotton in the Eastern Belt and receives a lower price.

The extent to which farmers realize profits from producing under these low costs is shown by detailed studies made by the Department in Lubbock County, Texas, for 1924.

> . . . The average net income per farm received for its operation by the family for the year, after paying all interest on borrowed capital (and including the value of that part of the family living furnished directly by the farm as part of the farm receipts) was for 139 farms over $3,000. Three men lost money, 10 per cent made less than $1,000, but 13.5 per cent made over $5,000 and the remainder made between $1,000 and $5,000. These incomes were obtained on farms averaging 232 acres in size, valued at $68 per acre, with 37.6 per cent of the land in harvested cotton having an average yield of 148 pounds of lint selling at an average price of about 20 cents per pound. The average net worth of these farmers when they settled in the region was just over $5,000 per man and on March 1, 1925, the corresponding figure was $18,000, the difference having been made by the operation of the farm and its own increase in value in an average period of operation of 5.84 years.[42]

The conditions in this county are typical of about eight others in the region.

Such an analysis serves to explain why the period from 1915 to 1926, which saw an increase in cotton acreage of 51 per cent in the whole United States, found the Western Belt with an increase of 87 per cent. Texas increased her acreage in this period 77 per cent and Oklahoma 169 per cent. The disastrous bumper crop of 1926 found several of the eastern states with acreage practically stationary and in some cases reduced. Texas and Oklahoma for that year increased production 2,076,000 bales, which almost accounted for the disastrous increase from 16,004,000 bales in 1925 to 18,618,000 in 1926. This analysis also serves to show why any state agreements to restrict acreage are likely to fail. When

[42] Wooten, *op. cit.*, p. 274.

other states are selling cotton below cost of production the newer regions continue to operate on a margin of profit.

Again Texas and Oklahoma have made cotton farming more hazardous for growers in other states because the variations in their yields are so great as to increase price fluctuations. Thus from 1925 to 1926 Texas with only a 4.2 per cent increase in acreage, from 17,608,000 to 18,363,000 acres, increased her crop 41 per cent, from 4,163,000 to 5,900,000 bales. In the same period Oklahoma with a 0.6 per cent decrease in acreage, from 5,214,000 to 4,912,000 acres, had a 15 per cent increase in production, from 1,691,000 to 1,950,000 bales.[43] These variations are due to the extreme danger of droughts in the West as already discussed. In this area the effects of the drought are increased by the blowing of the loose sandy soil which occurs during dry weather.

The crop of 1925 was grown under drought conditions; the season was ideal in 1926, and without perceptible increase in acreage over the preceding year the Western Belt pushed the cotton market to its lowest level since 1914. Such fluctuations defy cotton conventions and acreage restriction agreements; they are beyond the power of man to foresee and modify. The western farmers have thus unwillingly added to the hazards of cotton production both for themselves and for the growers who must compete with them.

When one turns from the fluctuations of cotton production and the cotton market to trace the effects of the cotton cycle on the South, he is likely to find that he has made a transition from economics to journalism.

---

[43] See "Cotton Production," Dept. of Agriculture *Yearbook*, 1926.

No one denies the influence of the cycle of cotton upon the rural and urban life of the cotton states. The tendencies seem obvious, but rural standards of living are illusive things, and it is only lately that they have received attention from the hands of social scientists. Until data can be accumulated for a realistic study of the fluctuations of the farmer's living with the fluctuation of the price of his commodities, a descriptive presentation from the reports of various observers is likely to be of some value.

Fluctuations in cotton affect the whole economic and social fabric of the South. Journalists do not betray the truth when they write:

> And what does cotton mean to the cotton states? It means life, health, happiness, and prosperity to them. In fact, nothing else matters much. If cotton is all right, all's well in the Cotton Belt. And if cotton is sick the whole South is sick. The physician can collect no bills, the merchant can sell nothing except on credit, railroads go without freight; mill operatives languish, children grow pale, every person in the street is dejected and gloom reigns throughout the South. . . . Cotton is the barometer that foretells the industrial fogs, squalls, and fair weather of the South.[44]

A good crop and a high price means more than that the farmer's wife can begin to dream of a new parlor carpet and a piano; it means that the preacher's son and the merchant's daughter can go away to college. The clerk scents a raise and cautiously enquires the price of a diamond ring for the girl that for the past two years he has been seeing home from church. The commercial traveler is lavish with more expensive cigars than he smoked last year, reflecting that the house

---

[44] E. V. Wilcox, "The Great White Way of Cotton," *Country Gentleman*, March 31, 1922, p. 78.

won't mind a bigger expense account, with orders coming in like this.[45]

It is known that tenants work up into land ownership in good cotton years; in poor years owners relapse into tenancy. Du Bois in a study of the Negro landholders of Georgia, made for the Department of Labor in 1901, pointed out this trend. After showing that his study indicated that Negro owners in fifty-four Georgia counties had increased their landholdings from 338,769 acres in 1874 to 1,075,073 in 1900, Katharine Coman points out:

> This threefold gain has been won by the most strenuous earning and saving. It means slow, difficult, patient achievement. Much depends on the price of cotton. In the years of high prices, 1884, 1890, 1900, there is a long step in advances, one hundred thousand acres being added each year. When cotton drops to five cents a pound, there is actual retrogression, as in 1894, 1898 and 1899. The cotton corner of the year just past [45a] was the means of putting thousands of Negro farmers in full possession of the land they tilled.[46]

Business becomes almost stagnant in certain cotton areas when prices fall. Farmers quit work and come to town not to buy but to stand around on street corners and talk things over. Cotton goes unpicked, partly because of a belief that this will cause the price to rise, but mainly because picking would add another expense to a crop already a dead loss. "I found many fields of

[45] Henry K. Webster, "Slaves of Cotton," *American Magazine*, July, 1906, p. 19.

[45a] The corner engineered by Dan Sully in 1903.

[46] "The Negro as Peasant Farmer," *American Statistical Association Publications*, IX (1904), 48.

cotton unpicked," wrote a traveler through the South in the late autumn of 1921, "some that would never be picked at all. In some cases it paid the tenant farmer better to desert his own crop and hire out to neighboring farmers to pick their crops." [47]

The rural South's almost complete dependence on a cotton economy was glaringly shown by the price debacle in 1914. The hardest hit were the plantation areas which were almost wholly dependent on the one crop. In the Alabama Black Belt in 1914, says a Bureau of Labor report, ". . . hundreds of landowners simply released their tenants from such contracts as they held against them. The rents were either relinquished outright or postponed indefinitely. . . . The customary advances of provisions to Negro tenants were cut off. Owners of large plantations were compelled for the first time in their lives to tell their Negroes that they could not feed them and that they were forced to let them move away. In a number of Black Belt counties food was distributed to the starving Negroes by the Federal Department of Agriculture and by the organization of the Red Cross. The tenants . . . were also in debt for provisions which had been furnished them during the past winter. Thus in many instances they lost their mules and other property which were taken for the payment of rent and store debts. . . . Nearly $50,000 was made up in and around the town of Demopolis, Alabama, and distributed among the most destitute." [48]

Studies on the subject cite the boll weevil and the low prices of 1914–15 as among the causes of the great

---

[47] Ray Stannard Baker, *New York Evening Post,* March 10, 1921.

[48] R. H. Lovall, et al., *Negro Migration in 1916–1917,* Dept. of Labor Bulletin, 1919, p. 61.

Negro migration. Now that the way out has been found for many people who never knew it before, migrations may be expected as the result of all periods of low prices. Dr. Andrew M. Soule, President of Georgia State Agricultural College, estimated that 100,000 of the farm population left Georgia during the first six months of the cotton depression of 1922. A survey carried on by county agents showed that 11,000 farms had been abandoned during that time.[49]

The reaction of war prices was in proportion. A greater amount of prosperity was distributed over the Cotton Belt down to the lowest levels of tenancy than ever before. A sympathetic observer from Georgia thus describes its interaction on the Black Belt:

> The Negro tenants are rolling in wealth. It is a curious sight, those in the country districts of Georgia report, to see them come to town with their pockets stuffed with paper money. They pay their bills and buy more goods with money peeled from great rolls of green and yellow backs. Theirs, indeed, is almost an embarrassment of riches. They seem hardly to know how to handle such masses of currency.
>
> One country storekeeper in Georgia states that since the last harvest he has taken in over twenty thousand dollars in cash that he had charged off the books. All this was for goods that had been sold in the past to Negro farm tenants. With prosperity suddenly come upon them, the Negroes made no attempt to avoid their just obligations.[50]

The war migration northward and the good prices for cotton combined to bring about more equitable conditions on the plantations and in landlord-tenant relations. Agri-

49 News item, *Raleigh News and Observer,* May 22, 1922.

50 Charles Lewis, "Thirty Cent Cotton and the Negro," *Illustrated World,* May, 1918, p. 470.

cultural labor, which had always been plentiful and cheap, rose to a higher economic level. In a letter to a newspaper a tenant described the new conditions for his class in the following terms:

> The tenant no longer canvasses a whole country in search of a home, for the landlord goes out into the byways and hedges and compels them to come. The tenant makes his demands as to additions, repairs to buildings in no uncertain terms. He says how much fertilizer he must have and the landlord gets it. He no longer plows a one-eyed stringhalted mule, for the landlord is on his note for a $700 pair.[51]

Standards of living were raised for farmers from North Carolina to Texas. Debts that had hung over men for years were paid. Storekeepers dug out forgotten accounts and presented them for payment. Land prices that had been stationary for decades rose almost over night. Writes Ray Stannard Baker of the rise and fall of cotton prices:

> The tremendous paying of old debts was harvest time for the merchants. In 1918 and 1919 . . . landowners and tenants had a taste of a better way of life, more of the common decencies and comforts, a better chance to educate their children, a little glimpse beyond the horizon of their dull ordinary lives. It is true there was great wastage . . . for example, a tremendous automobile invasion of the South, but everywhere I have found new houses, new barns, new roof paint, better roads, schoolhouses, churches, and better farm machinery. I visited certain neighborhoods where for the first time there have been glimpses of the comforts which are the common necessities of farm life in the North: home water supply, plumbing, proper lighting, washing machines, better furniture, musical instruments, telephones.

[51] *Raleigh News and Observer*, Oct. 6, 1921.

> . . . and now the curtain seems to be dropping upon that pleasant scene and they feel themselves headed back toward the old morass of crop lien farming—a kind of slavery to merchants and bankers.[52]

The relation of cotton prices to agrarian unrest in the South is clearly evident and interesting to trace. "The Tillman Movement" in South Carolina, as Francis Butler Simkins has shown in his book of that name, grew out of the inability of the upland cotton farmers in the coastal areas to make a decent living in the 1890's. Tillman himself failed at cotton farming and came first into the public eye by his vigorous demands for the relief of agriculture. Benjamin B. Kendrick[53] has held that the agrarian discontent in the South grew out of the crop lien system and the low social status of the southern farmer in 1880 as compared with 1860. But his low social status was to a large extent a low economic status owing to the falling prices of cotton since 1870. In North Carolina, Louisiana, Arkansas, Alabama, Mississippi, and especially in Georgia, the Agrarian Crusade took the form of a revolt against the Democratic party called Populism. Tom Watson came from a Georgia farm and was a country lawyer and farm owner before he became prominent. Mrs. Mary Elizabeth Lease had told her western audiences that what the farmers needed was to raise less corn and more hell! Down in Dixie, as some wag has accurately put it, they continued to do both. The black Republicans on the plantations raised the cotton, and the Populist Democrats, upland cotton farmers, raised the hell. The story of the cotton farmer at the

[52] In the *New York Evening Post,* March 8, 1921.

[53] "Agrarian Discontent in the South 1880–90," *Report of the American Historical Association,* 1920, pp. 267–72.

mercy of the market and falling prices, in the toils of the creditor, the supply merchant, the middleman, and the "anaconda" mortgage, going out to do battle against the solid South and the Democratic party, has been engagingly told for one southern state by A. M. Arnett.[54]

In the field of cotton economics the object of the farmers' particular wrath was the New York Cotton Exchange. It was believed that speculators established the prices ruinous to the cotton planters, and a particularly popular unwritten plank in the platform of the agrarian revolt was the extermination of the New York Cotton Exchange. A low price for a good crop of cotton was the result of the malign influence of speculators. The farmers were exploited. The attitude is well represented in an address delivered by Tom Watson at a cotton convention held during the crisis of 1905:

> What a singular situation is ours, my brethren. The world has never seen one similar to it. Famine has its millions of victims in India because the crops have failed. We sent the offerings of our charity a year ago to Russia because her crops had failed. Tonight in Ireland starvation clamors for its victims throughout the length and breadth of that afflicted country, and it is because they didn't make the crops.
>
> It is a curious state of affairs. I starve to death, not because I have no crust to eat, but because the table is bountifully spread. If we make no crop it is ruin, and if we do make one it is ruin too. It's the old predestination cry! You can and you can't, you will and you won't; you're damned if you do and damned if you don't.[55]

The speculators set the price of cotton. They lowered

[54] *The Populist Movement in Georgia,* chap. II.

[55] Cited by Henry K. Webster, "Slaves of Cotton," *American Magazine,* July, 1906, p. 305.

its price by creating cotton which existed in name only, cotton futures. Failing to exterminate the Cotton Exchange, Watson told the growers they had yet one weapon left:

> Spot cotton is king and always will be king if you will be true to it. The mills can't run on cotton futures. The gambling contracts made on Wall Street itself can't always be settled by other contracts. Sooner or later they have got to have spot cotton.
>
> What is Wall Street doing? It is simply betting that you will sell your cotton in April, May or June at six cents and a fraction. For them to win the game you have got to do it, for if you don't they lose the game and you win it. Put yourself with spot cotton just where Sully stood a year ago. Let us keep in our hands that which they are bound to have. Let us hold it like grim death and when they want it let them come to us and pay our price.[56]

The South has become used to political gestures in connection with cotton. When a sudden fall in the price of cotton has thrown the section into depression the appeal has been to oratory. Large numbers of legislative palliatives and nostrums are offered during each recurring depression. They serve as topics of conversation for a while; they bring their sponsors into brief prominence; but no one expects them to pass. "We've always been hollerin' for help from the Gov'ment," one farmer told Ray Stannard Baker, "but I've yet to find any man who can show me where we ever got it."

The agricultural surplus furnishes the most vexing problem in farm economic policy with which the nation has had to deal. The failure of laissez faire to minister

---

[56] *Loc. cit.*

to the welfare of the cotton grower has led to suggestions that include valorization of cotton, government subsidies to cotton growers, supervision of acreage planted, and suppression of crop reporting. A lack of clear thinking prevails.

The following résumé of bills introduced in the United States Congress during the 1926 crisis shows the varying viewpoints as to what is the matter with cotton, and the remedies that should be adopted:

1. A bill introduced by Representative Marvin Jones of Texas would limit the reporting and forecasting statement of the Bureau of Crop Estimates and Bureau of the Census. The South, it was explained, is turning against the methods of crop reporting, and is growing tired of the government's "arbitrary meddling with cotton growing and cotton prices."

2. Senator Tom Heflin of Alabama introduced a bill to make every bid and offer on cotton available to everybody on equal terms. This bill embodies the opinion current among many southern farmers that underhand methods of Wall Street and trading in futures are somehow responsible for the fall in cotton prices.

3. Another bill introduced by Senator Heflin would provide for a new census of cotton acreage planted in 1926 before the Bureau could proceed with the new estimates.

4. Senator Mayfield of Texas proposed making surveys of grades and staple lengths of stock carried over in warehouses.

5. Senator Cameron of Arizona introduced a bill to pay ten cents a pound bounty to the grower of all cotton of 1⅛ inch staple—the kind grown in Arizona. This proposal represents the bonus idea in its frankest form.

6. A bill introduced by Representative William C. Lankford of Georgia proposed to set up government corporations to buy all cotton offered at less than 22 cents and hold until salable at 24 cents.

7. Senator Pat Harrison of Mississippi proposed that the Federal Farm Loan Board set up twelve Federal Intermediate Credit Banks to lend to coöperatives on farm products.

8. Representative Tom Connelly of Texas urged that $50,000 be appropriated for research to discover new uses for cotton in industry.[57]

On October 13 a cotton convention, participated in by the governors of fourteen cotton growing states was held at Memphis, Tennessee. In the call for the convention, issued to all the governors of the cotton states, Governor H. L. Whitfield of Mississippi said in part:

> The rapid decline of the price of cotton within the last few weeks has brought the price of cotton below the cost of production. Within this time the cotton states of the South have lost hundreds of millions of dollars. . . .
>
> For the South as a whole, the cotton crop is the basis of all other business, and the depression resulting from this rapid lowering of price is already becoming manifest in all forms of business. . . .
>
> Experience has demonstrated that it is only a time of great necessity that causes the people to quit the old systems of business and adopt new methods. In other words, necessity is the great spur to change. All sensible men realize that such a situation exists today and probably the psychological situation is such that our people can be brought together on some general plan for the proper handling and marketing of our

---

[57] Aaron Hardy Ulm, "Congress Falters as to Cotton or General Farm Relief," *Commerce and Finance*, Jan. 12, 1927, pp. 125–27.

great staple crop. For a hundred years the South has been producing cotton and during this period there has been no systematic general method for marketing this crop on which the welfare and prosperity of the South rest.

The time has come when the business interests of the South, not only for the protection of the farmer but the welfare of the general business of the South must, through organized effort, come to the realization of the situation.[58]

Governor Whitfield closed by suggesting five items for discussion:

1. Formation of pools in every state to take off the market a sufficient number of bales to equalize supply and demand.
2. Definite steps to reduce cotton acreage proportionately throughout the Cotton Belt.
3. Utilization of governmental and coöperative growers' agencies toward these ends.
4. Development of new attitudes in southern business toward the marketing of cotton.
5. Suggestion of other plans.

At the meeting of the convention a campaign was launched to reduce acreage and to form cotton-holding pools. The latter met with more success. Credit organizations were arranged to assist those who wished to hold their cotton off a ten-cent market. A voluntary pool, financed by bankers and merchants, was proposed to take four million bales off the market and hold it for two years if necessary. The effect on the stock exchange was apparent the next day when the price rose half a cent.

In acreage restriction the convention was unable to

[58] Proceedings of Cotton Convention held at Memphis, Tennessee, Oct. 13, 1926.

do more than pay the customary lip service to diversification. It was virtually admitted that no hope existed in a voluntary acreage restriction campaign. One view is that the more successful a program of sign-ups to reduce cotton planting appears, the more likely it is that farmers will break their pledges. Another group holds that such reduction campaigns if successful are only temporary and serve in the long run to increase fluctuations. They follow the reasoning previously presented, that successful acreage reduction is accelerated by weather conditions to reduce production in greater proportion, and this so raises the price of cotton that next year all caution in regard to planting is thrown to the winds. Both views point to one conclusion—the futility of attempts at acreage reduction, as heretofore practised, to stabilize prices.

Much resentment was felt by those at the convention toward Texas for her disastrous increases in acreage, and the Texas delegates were properly repentant. The most daring and possibly the most hopeful plan, that suggested by A. G. Little of Blytheville, Arkansas, was defeated: It was based on the system in use in Egypt and called for identical legislation by all the states to restrict by law any farmer from planting more than half his land in cotton. The proposal was supported by delegates from Arkansas, Louisiana, Mississippi and Texas, strange to say, but defeated by the votes of the Eastern Belt plus Arizona. It was argued against the resolution that such a course would deprive the South of its monopoly on the world's cotton supply. "This may be one brand of political economy," wrote E. E. Miller, Editor of the Southern Agriculturist, "but a political economy of common sense would see that a country cannot be enriched by producing any commodity to sell abroad at prices that will keep the producers poor, and that it is better to lose a market monopoly than to hold it at the cost of starvation wages and low living standards."[59]

It is doubtful if such a use of the police power of the state would have been held constitutional. Compulsory diversification, if enacted, might have started a new era in southern agriculture. Failing that, the convention turned to the customary plea for reduction of cotton planting.[60]

Writing of the convention's failure to reach agreement, G. L. Fossick said of voluntary reduction of cotton planting:

> The idea is nothing new. The cotton South has often met fact with expedient and each time demonstrated that hope triumphs over experience. . . . Intermittently, for almost a century the cotton farmer has again been guilty of over-production, and each time he . . . has been offered the same nostrums.[61]

J. S. Wannamaker, president of the American Cotton Association, has suggested a plan for stabilizing the price of raw cotton which calls for neither national control nor voluntary coöperation of individual growers.[62] His plan calls for the appointment of a permanent cotton commission, composed of representatives from all the cotton states to study, advise, and prepare uniform legis-

---

[59] "Cotton a National Crop," *Review of Reviews*, July, 1926, p. 67.

[60] George Fort Milton, "Can Cotton Be Controlled by Law?" *Independent*, CXVII (1926), 531–32.

[61] "Ailing King Cotton," *Independent*, CXVII (1926), 380–82.

[62] "Stabilization of Cotton Prices," *Manufacturers' Record*, March 11, 1926, p. 86.

lation regarding cotton. The passage of uniform laws jointly by all the cotton states would then be sought in order to carry out the following program:

1. Forbid the planting of more than 40 per cent of any farm in cotton, and set up boards to have supervision and control of food, forage, and cotton acreage.
2. The passage of uniform laws requiring that a crop mortgage to be collectable must cover a certain per cent of food and forage crops in proportion to cotton.
3. The creation of financial machinery to retire such proportion of bumper crops as is in excess of demand to hold it for the lean years.

The first provision is in line with the laws passed by the Egyptian Government,[63] December 8, 1926, which provides that no farmer shall plant more than one-third of his tillable land in cotton on penalty of fine, imprisonment, and uprooting of the excess at the expense of the farmer. The fertility of the Nile lands has notably decreased since the building of the great irrigation dams, and the purpose of this legislation is to restore them by crop rotation and diversification. The second provision is suggested simply to give point to the first. It is thought that with the acreage stabilized by the first provision the third will simply take up and distribute the slack in production due to variations in the weather.

The difficulties in the way of the adoption of the plan, however admirable, are such as to pronounce it Utopian. Regardless of the slight likelihood of the cotton states coming to agreement, similar laws in America would be declared an unconstitutional use of the state's police

[63] Press Release, Dept. of Agriculture, March 26, 1927.

power. They would, if enforced, be likely to lead to diplomatic complications, as did Brazil's valorization of coffee or England's control over rubber. If the doctrine of comparative advantage holds in cotton, its enforcement would prove unjust discrimination against owners of rich Delta and alluvial lands especially suitable for the growing of long staple cotton.

It does not seem likely that any unified organization or control will ever be worked out that will enable the cotton industry to control to any extent its own destiny. Here, as elsewhere, in our competitive system men striving everywhere for their own best individual interest have at times worked themselves collective ruin in the process. Lack of knowledge, lack of common interests, lack of plan and organization have combined with the vagaries of nature and nature's insects to keep the cotton industry untrammelled by regulation.

Our survey will likely lead us to agree with A. W. McLean, former director of the War Finance Corporation and Governor of North Carolina, that "at the present time the cotton industry is the most hazardous of all branches of agriculture." It is not a matter of low average price alone, it is the matter of violent fluctuations which take place in the commodity within very short periods that make cotton dangerous. These fluctuations create hazards which must be assumed by all those who grow cotton. The cotton dealer and the manufacturer realize these risks and hedge against them. The grower cannot hedge his cotton. He bears the ultimate loss. It is these fluctuations that help to explain how cotton has brought the South collectively great wealth yet left the individual growers poor.

CHAPTER VI

AROUND THE YEAR WITH COTTON GROWERS

THE INDUSTRIAL routine of any group of workers forms the skeleton on which the living flesh of their social life is developed. To the diurnal, weekly, seasonal, and annual cycles, the community, civic, recreational, and even familial routine of the worker must conform. Thus a highly informing study of any human group would be found in charting its diurnal rounds. The typical daily record of a factory worker, an office girl, and a modern business executive would furnish significant data for sociological analysis. An annual record of the routine of the agricultural worker, a wheat farmer, a cattle ranchman, a truck grower, would be pertinent in that it would show variations in economic and social life due to the seasons. In his *The Peasants: Autumn, Winter, Spring, and Summer,* Reymont has depicted this cycle of the routines of Polish folk life with the graphic method of fiction. Such a study, it must be admitted, possesses descriptive value. It serves to enfold the abstract and generalized terms by which the economist describes the industrial processes with the warm reality of the cultural and social practices which accompany them.

It is suggested, accordingly, that a study of the men who tend cotton should also be a record of days and works as interrelated with cycles of the seasons. This chapter deals, then, with the routine, the movements, and processes of the human factors in cotton. The patterns for these processes are set by nature; they are cultural in the sense that they are determined by the demands of plants for cultivation; they are also cultural in the broader sense of social culture that has resulted from the adaptation of men to land and has been handed down as patterns to be followed.

If the man from Mars could obtain a bird's-eye view of the agricultural workers of the South as they move around the seasons with King Cotton he would observe a vast and ever changing panorama. In cotton fields over ten million men, women, and children each season are to draw most of their sustenance from the fleece of the cotton boll. He would see in many parts of the Cotton Belt men who from year to year seemed constantly moving about in aimless circles; in other parts he would look down upon men whose habitations were fixed. He would find black men and white men going around the season with cotton, Negro croppers in the deltas and the Black Belts of Mississippi and Alabama, working for white planters on great plantations; white tenants interspersed with black in the East; owners of small farms in red upland cotton areas. In Texas he would see a great new expanse of prairie cotton land coming more and more to be cultivated by white tenants. And as the cotton plant grows, flowers, and fruits he would note the human forces of many regions, men, women, and children, mobilized in cycles of movement that synchronize with the seasons and the plant.

The cotton year begins with January. By Christmas the cotton has been sold, the landlord and tenant have settled, the fertilizer has been paid for, and the supply merchant has balanced his bill. If the season has been a failure, the tenant may be informed that he owes, say, fifty dollars on his next year's crop. If the tenant already stands restless under the weight of failure, he may be driven to attempt to cancel his debt by moving out of the reach of the landlord or merchant. He may carry the debt to a new landlord who agrees to charge it to the unmade crop. Sometimes landlords wishing to keep good workers who have failed through no fault of their own wipe out the balance of debt and begin the year anew. At other times book credits seem so to keep themselves that the cotton worker, unlearned at accounts, never quite gets out of debt and remains bound to the land and a hard taskmaster. Many a harassed, hard-working Negro family must have gained its first release from debt in the migrations of 1916–21.

It is in December that the tenant in the Cotton Belt "takes a mind to move." The small owners stay with their places from season to season. But the last of December is for the cropper and tenant a time for deals. Over the neighborhood the tenant rides, holding conferences with this and that landlord about a place. A favorite meeting place is the small town on Saturday. Landlords looking for good tenants and tenants looking for good landlords hold conversations in country stores, at crossroads, in backyards, and on front porches—wherever they happen to meet. The landlord's questions are likely to deal with the crop made under the renter's former landlord, his reasons for leaving, the number in the family able to work, the acreage to be planted in cotton. The tenant wants to know the kind of house on the farm and the details of supplies to be furnished. The deal may be closed verbally or a lease may be signed. If the renter is satisfied that he can better himself by moving, he agrees to terms and announces to the family on his return home that next year "we are going to live on the old Brockton place over across the creek."

The mobility of the restless, roving cotton grower has become a proverb. Dr. E. C. Branson says of these tenants, "They move from pillar to post from year to year. They are a migratory type of farmer. They are cursed with the restless foot of the wandering Jew."[1] After the cotton harvest is over, one meets them in every nook and corner of the rural South, driving along the country roads with their scanty household goods piled in wagons, painfully exposed to the gaze of an indifferent world. In the United States it was estimated in 1922 that 19 per cent of all farms changed occupants. In eight cotton states from 30 to 40 per cent of the rented farms changed their tenants, the average for the whole South being 32 per cent. Less shifting occurs among the Negroes on plantations than for white tenants. The 1910 Census showed that Negro tenants had longer average periods of occupancy than white farmers, exceeding them from a third of a year to a year and a half for different tenure classes.[2]

Why do cotton renters move? They are driven by a restless search for something better. It may be they desire a better house, a flower plot, a better school. Less likely they have fallen out with a landlord, and with a strange mixture of pride and inferiority they seek to show their independence by moving. Sometimes they move secretly and with no warning in order to leave him in the lurch for some slight, real or fancied. Most of all they

[1] *Home and Farm Ownership*, North Carolina Club *Yearbook*, 1921–22, p. 74.

[2] L. C. Gray, et al., "Farm Ownership and Tenancy," Dept. of Agriculture *Yearbook*, 1923, p. 592.

move because they believe that just over here they can make better crops, "get a holt of some money," "come out better'n we did last year." The Department estimated in a study of reasons assigned for shifts by 3,360 tenants in Kentucky, Tennessee, and Texas that 20 per cent of the moves indicated progress up the tenure ladder for renters and croppers. Partial or complete crop failure accounted for 14 per cent of moves for croppers, and 9 per cent for share tenants. The desire to obtain a farm better adapted in size, soil, and improvements was assigned as reason for moving by 25 per cent of the shifting croppers and 31 per cent of the share tenants.[3]

Elizabeth Madox Roberts has performed a service to art and truth in *The Time of Man* by her presentation of the tenant as forever moving in search of the unrealized yet to be attained. Like many of his more fortunate brothers, the tenant lives on hope with his daily bread. It may be that moving is the one luxury that comes in the dull and monotonous round of the tenant life. "Moving day means a different thing to every member of the family, but on the mothers the most burdensome part of it falls. To them it means the labor of taking down and packing up in the wagons all the family possessions, . . . traveling to the new place of existence, cleaning, scrubbing, and placing the household goods."[4] To the children there is nothing quite so thrilling as moving day. Miss Roberts, writing from first-hand knowledge, describes convincingly the pleasure of the tenant child at exploring each new house to which the family comes. If so, it is a cheap luxury, and who shall begrudge it to their hard lives?

---

[3] *Ibid.*, pp. 595–96.

[4] Lucy C. Crisp, *Raleigh News and Observer*, Jan. 10, 1926.

After renting the land the tenant has to see about getting furnished for the year. If a share tenant, he has his own work stock and tools; if a cropper, he has nothing. On a delta plantation his supplies would be furnished by the landlord. The landlord, however, may agree to stand for him at a supply store. Small owners unable to keep themselves during the growing season also apply to these stores. In such cases goods are furnished on open book credit at credit prices which may reach 25 per cent. Very often the grower may be limited to drawing a certain sum, say $10 or $15 a month. In emergencies he may secure an increase by special pleading with his landlord. Crises in the family, such as illness, play havoc with these makeshift budgets. The families tide it over as best they may. Either the doctor goes uncalled or realizes that he must wait until fall for his fee. Oftentimes he never gets it. The services rendered for sweet charity's sake by harassed country doctors in the South have not received adequate recognition.

The merchant and the planter must also be financed. The southern planter once received advances from his factor. It is safe to say that most plantation owners preparing to plant a crop now go to see their bankers. After a conference covering details of amount of acreage, number of plows, costs, and production, the planter is given a line of credit secured by personal note. The loans are to be issued according to seasonal needs; so much for furnishing, planting, chopping, and picking. The supply stores are granted credit by wholesale houses. This, however, is considered precarious business since the failures of 1914, 1921, and 1926. Accordingly, many supply merchants now borrow from banks to pay off their thirty-day bills with the wholesaler.

To the landlord and tenant, the large planter and small farmer the question of credit is preliminary. After supplies, work stock, tools, land, and men have been brought together, they await the turn of the season. "Christmas" lasts a long time. From the close of cotton picking to the beginning of cotton planting includes a vacation of two to three months. The man who grows cotton and corn finds little to occupy his time except chores and attention to the stock. Except for vegetables in the extreme South, no crops are grown to use the farmer's labor during this period. The family is likely to consider itself lucky if the men-folks during these winter months are able to secure employment on road work or levee repairing. For the delta plantation owner it is a period of suspense and uneasiness only to be relieved when the Negroes go to work. The black boys drive about the country roads at a great rate in their second-hand cars, go possum and rabbit hunting, sometimes drink too much, and once in a while get into serious cutting affrays. The orthodox southern planter is likely to feel that his tenants are wasting their substance in riotous living. He is torn between the feeling that some of them ought to be in jail for tearing around the country so, and the devout hope that none of them will get into trouble with the officers. He may make no effort to put a stop to their hilarities, but he heaves a sigh of relief when the routine begins.

Professor R. P. Brooks, writing from first-hand observation of plantations in the Black Belt of Georgia, is rather severe:

> Tenants usually spend all of the proceeds of the year's work before Christmas and return to the landlord for small sums to tide them over the holiday season. The writer cannot recall an instance of a planter who had not found it

necessary to make such advances. . . . As a general thing the [tenant's] surplus melts away without substantial return. Whiskey, gambling, indulgence in sexual pleasures, purchase of useless articles of luxury, and excursions to distant towns, absorb their profits. . . . Christmas money is commonly advanced on the agreement of the tenant to renew his contract for another year. It is said that the landlord who would refuse to make these advances would very likely be unable to secure tenants.[5]

It is doubtful if the white tenants enjoy the vacation so much. Hunting, driving, loafing about the small town stores are regular pastimes with them. But puritan strains have rendered their pleasures less simple and direct than those of the Negroes. In many places the merry square dances once prevailed. The rural preachers have ruled dancing out. The southern cotton farmer does not know how to play. His recreations are little mixed with the strains of culture, and he is likely to grasp pleasure with a crude and brutal hand. Before prohibition the orthodox dreaded Christmas in the little inland towns of Arkansas, Mississippi, and Georgia. Drinking sprees and shooting affrays were common during the winter months. With the passing of time, manners and customs have softened somewhat, but the rural distiller and bootlegger plies a brisk if less obvious trade. "If cotton prices are low," the saying goes, "there will be a quiet Christmas this year."

The fields are left barren from the last year's harvest with black and withered cotton stalks straggling down the old rows. The farmer's first task is to cut or break the old stalks. He may clear the land in January or Febru-

[5] *The Agrarian Revolution in Georgia, 1865–1912,* p. 99.

ary unless he has burned the stalks just after harvest in an attempt to prevent the hibernation of the weevil. At the same period the corn stalks if left in the field can be cut. The disposal of the old stalk is often left until planting time. The farmer is able at the usual rate to clear about seven and a half acres a day. The stalks may be turned under. It is becoming more the practice, especially in the Western Belt, for the farmer to plow up the stalks, rake them up with a hay rake, and burn them.[6]

Cotton planting begins about the middle of March in southern Texas and moves upward through the belt to reach North Carolina and the northern margin about April 21. By the first of April, cultivation is under way in the Black Waxy Prairie of Texas and in central belts of Louisiana, Alabama, and Georgia.[7]

In order come the processes of plowing, bedding, harrowing, and planting the fields. The following description refers to the cultural practices in the Eastern Belt:

> A one-horse plow is run along the side of the old row and the dirt is thrown toward the middle by running one or two furrows on each side of the row. This leaves the old cotton stubs standing on a balk or small ridge, which is broken out with a one-horse shovel plow, leaving a furrow in which the fertilizer is distributed with a one-horse fertilizer distributor. The land is bedded back on the fertilizer with a plow taking from two to four trips per row. This leaves another "balk" between the two old rows, which may be left until the cotton is cultivated. The top of the bed is leveled off with a harrow or a board and the cotton planted

[6] *Cotton Altas,* p. 15.

[7] Baker, *Seedtime and Harvest,* Dept. of Agriculture Circular 183, p. 36.

on top of the bed, the new row being in the same place as the old one.[8]

This practice gives the new cotton plants the left-over benefits of last year's fertilizer. The rows are left about three to four feet apart. The farmer is likely to use a bushel of seed to the acre. On many farms it is customary to drop the seed and fertilizer at the same time, using a combination planter and distributer. The whole operation, with the use of one-horse implements, requires almost a day to the acre. It is the very opposite of time and labor saving. The small size of the eastern farms and the ruggedness of their contour, however, have served to prevent the general use of better methods. At the same time the farmer must plow and lay off rows for his corn. These operations easily require more than a day's work per acre. The supplementary crops of cotton and corn thus interfere rather seriously with each other.

In the flat alluvial areas and especially in the western prairies, the level lands make possible more efficient cultivation. Little fertilizer is used and the processes depend less upon man labor and more on machines and mules. The following description is typical of practices in a Texas Black Land county:

The beds are thrown up with a four-horse "middle-buster," the mouldboard of which resembles the mouldboard of a right-hand and left-hand turnplow fastened together. This is run in the old row and the dirt thrown to the middle on either side, makes beds in the middle between the old rows. When the beds are completed the land is harrowed. . . . The cotton is drilled on top of the bed, one row at a time

[8] *Cotton Atlas,* p. 13. The crop practices are those found in Anderson County, South Carolina.

with a two-horse planter which both smooths the top of the bed and drops the seed.[9]

As in the East, these rows are from three to four feet apart and require about a bushel of seed to the acre.

The contrast between upland and prairie land cotton culture is essentially that between a walking and riding cultivation. In the East the "cotton growing syndicate," in Tompkin's phrase, is a Negro behind a mule and a Georgia stock; in the West it is a white farmer riding a two to four-mule cultivator. Records of the Department show that the planting process for upland lands in Georgia, Alabama, Mississippi, Louisiana, and Arkansas requires fourteen to eighteen hours of man labor plus an average of twenty-two hours of mule labor per acre in cotton. In the Texas Black Land the farmer spends five hours of work to fifteen hours for his horse per acre.[10]

Cotton planting is general throughout the whole belt during April. It ends usually by May 21 in the northern areas, the process requiring about a month. Given the right amount of moisture, spring warmth, and no frost, the cotton farmer has not long to wait for the little plants to show just above the ground. The bare clean fields with the rows of tiny two-leaved plants stretching away into the distance are a sight to gladden the heart of any veteran cotton planter.

As soon as the small plants are up to a stand, about a month after planting, the farmer begins to "chop out" his cotton. The first cotton chopping begins in Texas May 1 and reaches North Carolina by May 21. The plants are thinned out to furnish better growing condi-

[9] *Ibid.*, p. 14. The description is of Ellis County, Texas.

[10] Baker, *op. cit.*, pp. 36, 37.

tions. The old plantation rule of "once a week and one in a row" sums up current practices. The working in of fertilizer also produces better results. Hoeing cotton forbids the use of machinery to any great extent and must be done by hand. Next to picking it is the most laborious process in the cultivation. Except in the western area such weeds as crab grass and Johnson grass grow rank throughout the South. After chopping out, which lasts from four to five weeks, it may be necessary to continue cultivation. In the East, chopping cotton requires from fifteen to twenty-five hours of farmer's labor per acre. Texas, however, on the average demands only eleven hours of work for the same process.[11] If he cultivates by plow, the farmer must avoid deep plowing alongside the plant, for this practice causes the early bolls to shed. The average cotton farmer has one answer to the cotton plant's demand for hand tillage, and that is to put his wife and children in the field. If he were able to pay the bills he would find but little agricultural labor to hire, for all his neighbors are busy chopping their own cotton. Except for Mexicans in Texas there have developed no migratory laborers in cotton as in wheat. The cultivation of his corn acreage requires almost two and a half days per acre and comes at the time cotton most needs attention. Accordingly, the farmer drafts his family. It is true that the women and children, white and black, of the small cotton growers are more accustomed to work in the fields than those of any other farming group in the United States. Travelers through the South for the first time carry away vivid memories of tired women, leaning on cotton hoes, staring at the passing train. By those reared

[11] *Ibid.*, p. 38.

in the South, women and children working in the field come to be accepted as part of the natural order of things. The women and children carry their hoes and files out to the field; the mother leaves the baby on a "pallet" under a shade tree, and occasionally stops to nurse it. Just an hour before dinner she leaves for the house to prepare a "snack." After dinner the family returns to the field and works until dark. This description applies equally well to large groups of black and white farm families.

It is hard to say just when the crop is ready to lay by, because the farmer never knows the amount of cultivation his crop is going to require. A wet April or May will keep the weeds growing as fast as the cotton plants, and one chopping is hardly finished before another is necessary to "get the cotton out of the grass." Continuous cultivation forces the growth of the cotton. Since the advent of the boll weevil it has accordingly become the practice in many localities to continue cultivation "till the last moment in order to keep the plant making as many squares as possible." [12] Since weevils attack them first, the late squares serve as a protection to the bolls already formed.

In normal years the crop has begun to look well by June. Everybody expects a good yield. The Department's forecasts are likely to be high. Hubbard cites an old saying of the cotton trade that "The man who can once sell the June prospect at the August price will never have to work again." [13] By the middle of July the cotton plant has begun its fruitage. First appear the squares, three leaves folded together with the bud inside. Given fair

[12] Hubbard, *Cotton and the Cotton Market,* p. 29.

[13] *Ibid.,* p. 31.

weather, it is three weeks until the squares open into cotton blooms. The blossom resembles that of the hollyhock, first cream colored, next pink and red. On the third day the bloom withers and drops to the ground, leaving a little pod. If the southern sun shines on, the folk saying has it right: "It takes six weeks from the bloom to the grown boll."

The routine of the cotton cropper and tenant is incomplete without an account of the activities of the landlord, the plantation owner, and the supply merchant during the growing season. In general, it is safe to state that the landlord living in town visits the tenant when he fears the crop is being neglected; the tenant visits the landlord when he wants to ask for something, be it wire screens on the window or more seed for planting. The tenant's every trip to town on Saturday is occasion for a conference. But if the trips become too frequent the landlord is likely to suggest that he is neglecting the crop. Sunday is occasion for the vigilant landlord to visit his farm, walk over his acres, and inspect the crop. The tour often ends with a visit to the tenant's shack and much good advice. The advent of the boll weevil caught tenants and croppers unprepared, and the landlords and managers have supervised the fight in many cases.

The plantation manager or owner keeps a closer tab on his renters than the town landlord. Negro croppers in the Black Belts and the Deltas have more of the status of hired men paid with a share of the crop than of tenants. They are under close supervision, and on some plantations the work stock is returned to central farms at the close of the day to be fed and secured from thence in the morning. The routine in planting, fertilizing, and chopping is standardized, and each cropper and his family are instructed by the manager to follow a set routine. While this practice in many instances makes for efficiency, it is likely to go against the grain with many tenants. Many of the white renters, for instance, are, with the exception of good advice, left to shift for themselves.

The supply merchant also exercises an amount of supervision over "his farmers." Each Saturday trip to town is an occasion for discussions of weather, markets, and weevil, as well as for purchases of side meat, sugar, flour, and cheap coffee. The relations are social as well as economic, and it has been observed that the merchant bestows more attentions upon his debtors than upon the cash customers. Rough and ready jests mixed with reminders of a mortgage on a certain worthless mule, presumably blind, are exchanged. The Negroes seem to have an uncanny sense of how far they dare to go in jest, and many a loud laugh and "Yass sir, boss, that's sho' the truth" enliven the country and small town store on Saturday.

Cotton is usually laid by about the first of August. This means that cultivation has put the plant far ahead of the weeds, and the farmer must now wait for the chemistry of sun and soil to do its work. His hopes are for dry hot days with moist hot nights, and in normal times his hopes are answered, for this is the southern climate that has made cotton a southern crop. It is almost a month until cotton picking begins, no other crops are ready for harvest, and the cotton workers are free for a time. In the country districts the summer term of school may be in session, for when autumn comes the children will be picking cotton. It is after the cotton is laid by that protracted meetings are held all over the South.

In lowlands and uplands, strange sects, the Holiness, the Nazarenes, the Holy Rollers, draw the croppers and their families to the bush arbors, or maybe the family packs up and goes to the camp grounds for several weeks. There the shining lights of the sect are gathered and rotate in preaching their best sermons. The summer revivals partake of the nature of a southern culture trait. None is more religious than the Negro, and late into the summer nights in many a countryside can be heard the sweet strains of an old spiritual and the rhythmic rise and fall of the preacher's voice announcing judgment to come. The landlords and town dwellers are likely to belong to the more orthodox sects, but they, too, have their revivals during the lull in the cotton season.

There is much dissatisfaction among landlords with the habit among cotton renters of easing up during August, "but it seems sensible to me," writes one.[14] "Cotton anyway must be grown by a series of spurts rather than by a steady daily grind." During the hot season "about the best thing for croppers to do is to quit work, visit around, and attend the protracted meeting. Then if they haven't killed each other *ad interim*, they are physically fit when the rush of cotton picking begins."

H. C. Brearley found in a study of 1601 homicides in South Carolina from 1920 to 1924 that "months of high homicide rates concur rather closely with the seasons of little farming activity, with one peak during the winter vacation and the other during the midsummer lay-by and camp meeting time. Two of the three months of least

---

[14] Alfred G. Smith, "The Cropper System," *Country Gentleman*, Sept. 4, 1920.

homicides, May and October, are also the months when farm labor is most busy."[14a]

By the first of August cotton picking is well under way in south Texas. It moves up the belt until by September 11 the whole rural South is engaged in harvesting its cotton. The bolls have opened, and the white fluffy fibre stands encapped in the five pointed star of the opened boll—Dixie's trademark. The harvest is to continue for three months. The bolls open at different times, and the fields must be picked over thrice or more. Cotton picking requires more hand labor than any other process in contemporary agriculture. It has never yet been successfully mechanized. The average Negro hand is expected to pick 150 pounds of seed cotton a day. In the Western Belt, where the bolls grow larger and open wider, the average adult picks 250 pounds a day. Depending on the yield it takes a man from thirty to a hundred hours to pick an acre of cotton. The average time required outside Texas is estimated at fifty hours of man labor an acre. Seventy-two planters of Louisiana averaged over six and a half days per cotton acre in picking and hauling the cotton.[15] The amount of the farmer's time required to pick an acre of cotton would produce three acres of corn in Iowa or four acres of wheat in Kansas.[16]

At the approach of the picking season the rural South begins to mobilize its labor force. Business picks up. Negroes from the towns and villages load on the planter's wagon or truck and journey out to the fields. Maids, cooks, and men-of-all-work desert their employers with

---

[14a] *Homicides in South Carolina, 1920–1924*, p. 22. An unpublished study, University of North Carolina, 1928. South Carolina had a population 82.5 per cent rural in 1919.

[15] *Cotton Atlas*, p. 15. [16] Baker, *op cit.*, p. 39.

mingled announcement and apology to make "cash money." It is a saying in the South that it takes a country Negro seven years to forget the call of the cotton fields. Loud is the grumbling, and many are the aspersions cast on the loyalty and honesty of the family cook, but she is likely to continue to accept her chance to make the highest wage offered the southern agricultural laborer. The demands of the cotton system are inexorable, and the townspeople continue to bow to them. Payment is likely to range from 75¢ to $1.50 a hundred pounds, and a good picker may make as high as $4.00 a day. Lower wages are paid for picking cotton earlier in the season, and the price may advance from a dollar to $1.25 a hundred. There is a reason. The first cotton to open is heavier, since it is still moist from the boll, and the leaves prevent drying. Later the fibres, being exposed longer, dry out, and a higher price rate compensates for their decreased weight.

Whatever it may be to the worker, the cotton fields at picking time have become to the world a symbol of Dixie. The benevolent blackness of an old mammy's face, the brightness of her dress, and the perspective of white cotton rows blend in colors dear to the hearts of souvenir buyers. It has become the sentimental postal card picture of the South. It is a symbol in the sense that the plantation was a symbol. In far-away lands men's hearts have been known to lift with joy at the sight, as they do at the sound of Dixie. It is noteworthy that these postal cards do not picture white women and children as picking the cotton. Cotton picking is not all festival. The sun is warm, and the Negroes laugh and sing, but Negroes have been given the tradition of laughing and singing at notably hard lives. Henry W. Grady pointed out long ago that the Negro spent his life with cotton and yet he had developed no folklore and no songs around the plant. Since then the Negro has sung of Mr. Boll Weevil a-sittin' everywhere. One wonders after all if the postal card picture is not a symbol, the symbol of a feudal order and of the domination of a section and a race by a plant.

On the rented farms the tenants and croppers must pick their own crop or pay the labor bill. Consequently, they and the small owners draft their wives and children. All go to the fields early and work late, always ten hours, and often twelve a day. Each picker wears a long sack which fastens over the shoulder by a cord and drags behind him on the ground. With bending back and both hands flying from boll to boll, the picker takes one row or two at a time. As his hand becomes filled with cotton he thrusts it into the mouth of the sack behind him. When he is working for wages, the picker carries the filled sack to the weigher, who hoists it on a steelyard and notes the result in his book. The cotton is poured into an unhitched wagon with high sideboards and tramped down. If damp with the morning dew it is emptied on a wagon sheet on the ground to dry before weighing. On the large plantation the fibre is stored after each day's work in the little cotton houses which dot the landscape. The white tenants and small owners are likely to dump the cotton on their small front porches until they are ready to haul it to the gin. A wagon load of seed cotton, "tramped down" between the high sideboards, makes a bale of lint.

The haul to the gin is not likely to be long. It is estimated that the time required to haul the crop to gin and to market is four hours for the farmer and eight hours of mule labor per acre.[17] In the real Cotton Belt there is a gin in every country town and at many country crossroads. The farmers draw their wagons up in a line and wait their turn to drive under the suction pipe. During the ginning season a great amount of time is spent waiting on the gin, but for the busy farmer it is a time for sociability and gossip with his neighbors. When his turn comes the farmer drives under the shed and helps to move the flexible suction pipe about over the loose cotton. In a short time the fleece is sucked up. If he wants his seed back, he drives around to a chute where it is poured into the wagon. Very likely, however, he will sell the seed to the ginner at current market prices. The bale is delivered to the farmer at a cost ranging from $3.00 to $4.00 for ginning, bagging, and ties. On the plantation and rented farms tenants and landlords each pay for ginning their share of the cotton.

How the farmer sells his cotton depends on his tenure status. Every southern town, large or small, has its quota of cotton buyers. The different types of buyers found are cotton factors, supply merchants, landlords, street buyers, and buyers for cotton firms. "General Merchandise and Cotton Buyer" is written on the signs of many small commercial gentry of the Cotton Belt. The tenant either lets the landlord sell his cotton or may sell to him direct at current spot prices. The creditor may deliver his cotton to the supply merchant and receive the balance after paying his debts. The farm owner has a wider range of choice among the local buyers. The towns are scenes of hectic activity during the picking season. Farmers drive into town with wagons piled high with cotton

[17] Baker, *op. cit.*, p. 37.

bales. Each buyer mounts the wagon hub, takes a slash at the bagging, and tears out a generous handful of sample. When sampled by all prospective buyers, the bales present a mottled and a gapping appearance. Many ignorant farmers believe that buyers make a great deal of money by selling all their samples. If the first bids offered are not satisfactory, the farmer may take a sample and hunt up a merchant, who he thinks will pay a better price. If unsuccessful he may take the cotton home and hold it. Cotton is also deposited in warehouses, the staple certified, and warehouse receipts issued. These receipts are negotiable and can be used as security for loans up to about 80 per cent of the market value of the cotton.

The intervals between pickings over the field can be put to a good use by the cotton grower. "Corn pulling" time occurs in early November and takes almost a day per acre of the farmer's time. More arduous is fodder pulling, distasteful because of the stinging sensation the fuzz gives the sweating laborer on face, arms, and neck. This operation also requires about a day to the acre.

Demanding hard labor but always welcomed, hog killing is something of a festival to the inhabitants of the Cotton Belt. Hog killing waits on the first "real cold spell" when there is no likelihood of warm weather's causing the meat to spoil. Stunning, sticking, scalding, scraping, disemboweling, cutting up into joints, and salting down the fatted swine from the pen are operations likely to be done with neatness and dispatch by the farmer, assisted by his grown son or a neighbor. On the farm woman devolves the labor of rendering fat into lard, strips of fat and lean into sausage, and feet and head into head cheese or "souse." Mrs. Julia Peterkin in *Black*

April has truthfully and artistically portrayed the important rôle that hog killing plays in the life of the plantation Negro, but it is no less significant to the white family. Existing on a monotonous diet throughout the growing season, with little or no consumption of beef or mutton, the grower's family is "almost starved" for fresh meat. A butchering is a neighborhood event, and presents of the delicacies are distributed. First to be eaten are the brains and liver. Liver hash and liver pudding are special delicacies. It is commonly thought that only Negroes eat the "lights" (lungs) and "chitlings," but there are many white cropper and tenant families who regard them as palatable. Backbones, boiled, and spareribs, fried, follow. The crisp remains of fat after the lard is extracted are often baked in corn meal to make crackling bread, esteemed on almost every southern table. Hams, shoulders, and side meat are smoked and stored away to furnish meat throughout the winter.

It is likely that by this time the rural schools have opened for the winter term. If so, the attendance of children, both black and white, is likely to be sporadic until cotton picking is finished. School laws allow children to be excused for emergencies on the farm. Compulsory education is confronted by a fact, not a theory, and the school bows to the cotton system. In many places schools do not open until after the picking season. And until living standards and economic status of the cotton farmer are raised this seems to be a realistic adjustment.

After several more pickings the crop is finished, all the cotton has been baled, hauled down main street, and sold. If grown east of Texas, the crop has required from 100 to 140 hours of man labor and from 45 to 60 hours of mule labor per acre. The amount of labor has varied according to the soil, the yield, the climate, and current cultural practices. It is the time to settle up. With warehouse receipts in hand, the "general merchant and cotton buyer" goes over the books with his client, the cotton grower. The bills are paid, and if the season has been good the grower pockets the surplus in his overalls. The landlord and his tenants hold their accounting of so much furnished, so much rent, so many acres of corn, and so many bales of cotton. The merchant settles with his wholesale jobber and pays off his note at the bank. The landlord is also in position for the first time to visit his banker.

Already business has begun to grow brisk in the Cotton Belt towns. The drummers have made their rounds, scented good cotton crops, and received large-sized Christmas orders. The farmer's family now drives to town every Saturday, the "old woman" sits hunched up in the board seat with her lord and master, the kids and "younguns" are likely to be piled on the floor on quilts. The whole family is likely to be fitted for shoes; goods for new dresses for the wife and the "kids" are in order. The merchants do a rushing business and attempt to recoup in three months for the stagnation of a whole year. Cheese and crackers, salmon and sardines, ice cream cones, and bananas help to make Saturdays a treat for all. Late in the afternoon the family starts home, "all tired out" from walking the streets, but with more than their usual supply of flour, coffee, sugar, with possibly some "fancy groceries" included. While the parents were gossiping, the children spent their time looking into store windows. And they are not forgotten. "I got to buy some Christmas for the kids; they're kinder expecting it," is heard in the South regardless of the price of cotton. A doll, a knife, some fireworks, apples, oranges, nuts, and candy are the legitimate right of every child, and the cotton cropper's children will find them in their stockings Christmas morning.

The cotton grower's vacation has begun. Croppers on the Black Waxy of Texas, Negroes on the plantation in the Mississippi Delta, the small farm owners in the Georgia uplands have finished the routine of the seasons and added a billion dollar crop to the world's supplies. For a time they stop and talk about "having a good Christmas." After respite comes again the question to move or not to move. And soon again it is time to cut stalks and plow land. Another billion dollar crop is under way.

CHAPTER VII

THE COTTON SYSTEM AT THE TURN OF THE QUARTER CENTURY

The growing of cotton is characterized in extensive agricultural production in America by four interrelated factors: Cotton is produced with a large percentage of expensive credit; it is often produced in connection with a small percentage of other crops; it is marketed in a system over which growers have developed little control; and it calls for a great amount of manual labor, much of which is furnished by women and children. It is necessary to consider these four elements in the system by which cotton is produced.

In so far as the production of cotton may be regarded as organized, it is organized around the system by which it is financed. In this connection are included the plantation organization and the crop lien which have already been discussed. The sources of credit for cotton production, the number of farmers needing such credit, the nature, the amount, and the charges for the credit must be considered.

In the Delta areas[1] the plantation owners secure the means for growing the crop from at least four principal sources: banks, wholesale merchants, cotton factors, and

---

[1] W. J. Carson, *Financing the Production and Distribution of Cotton,* Federal Reserve Bulletin, 1923. The Delta has come to be a popular term referring to alluvial regions of the Mississippi Valley.

local merchants. The factorage business has recently suffered heavy losses, but in 1922 from 5 to 10 per cent of planters' advances were received from factors. From 40 to 80 per cent of these advances come from local banks. From wholesale merchants may be secured from about 10 to 20 per cent of the credit needed. Wholesale grocery companies operating in the Delta regions estimate that about 50 to 90 per cent of their business is with plantation commissaries. For croppers and share tenants the one source of credit is the landlord. Usually the planter gives the tenant a line of credit at the plantation supply store. The credit is based on the tenant's acreage planted to cotton and is secured by a crop lien. Each month the renter is allowed to draw part of this line of credit. The time price is usually 10 to 25 per cent higher than the cash price. In 1920 and 1921 the average amount of credit furnished to 1,330 [2] families of Negro cotton. growers on plantations was $289 to croppers and $555 to tenants. These advances are used by all the plantation labor regardless of the returns made from cotton. "Cash money is for Christmas; furnishing is for livin'," is the way one cropper phrased his attitude. The inability of the cropper to finance himself, in or out of the plantation system, is shown by his lack of capital. An estimate [3] of the per capita net worth of persons engaged in farming in the United States of January 1, 1920, assigned the following values: owner farmers, $13,476; part owners, $12,829; tenants, $4,315; and croppers, $354.

The problem of financing the cotton grower outside

---

[2] Brannen, Dept. of Agriculture Bulletin 1269, p. 62.

[3] L. C. Gray, "Accumulation of Wealth by Farmers," *Proceedings of American Economic Association,* March, 1923.

the plantation area is larger and more complicated.[4] The upland areas produce almost nine times as much cotton, and a greater variety of credit institutions exist than in the Delta. Small owners secure credit from banks or supply stores, while the landlords make a larger use of banks. The prevailing rate of interest charged by banks on personal and collateral loans to farmers in the leading cotton states in the spring of 1921 was found to range from 6.23 per cent in North Carolina to 9.70 in Arkansas.[5] The forms of security most used are, first, personal notes with one or more endorsements, second, mortgages on live stock and, third, crop liens. The cash and share tenants secure credit either from their landlords or from fertilizer dealers, supply merchants, or banks, usually with the landlord's aid. The croppers, the most helpless of the group, are usually forced to depend on landlords and local supply merchants. The survey of 1923 found that only 12 per cent of the cotton producers were able to finance the entire growing process themselves. From 50 to 90 per cent, depending upon the locality, "borrow in the spring and continue until the cotton is marketed." [6]

Except for bank credit, which ranges around 8.5 per cent, the interest charges to farmers during the cotton growing season are almost impossible to estimate. The difficulty lies in the fact that much of the bank credit comes to the growers at second-hand through fertilizer dealers, landlords, and supply stores. In North Carolina, South Carolina, southern Mississippi, and west Tennessee the growers secure a little more than half their credit

[4] Carson, *op. cit.*, pp. 17–27.

[5] Dept. of Agriculture *Yearbook*, 1921, pp. 368–69.

[6] Carson, *op. cit.*, p. 18.

from supply stores, while Arkansas and northern Mississippi report that about 75 per cent of the stores sales are for fall payment.[7] Although the amount of credit extended by the supply merchants to the tenant varies from year to year, it is a very small part of their total credit needs. Croppers lack sufficient chattels for security for either the banks or supply stores to furnish them, and are often supplied entirely by the landlord. "It is safe to assume that the landlord furnishes more than half the credit needs of the tenants." [8]

Various studies of the cost of furnishing by supply stores have shown high interest rates. Clarence H. Poe in a study of the credit system of ten cotton states estimated that time prices averaged 23.33 per cent above cash prices. If time prices on an average carry over four months he says "it is equivalent to paying interest at the rate of 70 per cent per annum." [9] Studies in North Carolina farm credit have shown lower interest rates. One survey of 800 farms in three North Carolina counties [10] showed that 54.2 per cent of the farmers used merchant credit for which they paid 26.6 per cent interest. It was found that 40 per cent of these advances were fully secured by crop liens. R. B. Eutsler found that out of 588 Negro farmers 52 per cent used merchant credit for which they paid an average interest rate of approximately 26 per cent.[11]

---

[7] *Ibid.*, pp. 18–19. [8] *Loc. cit.*

[9] See news report of Ray Stannard Baker in *New York Evening Post*, March 10, 1921.

[10] F. R. Yoder, H. S. Beardsley, and A. J. Honeycutt, *Farm Credit Conditions in North Carolina*, North Carolina State College Bulletin, 1923, pp. 2, 21.

[11] *Negro Agricultural Credit Conditions in North Carolina*, MS., Institute for Research in Social Science, Univ. of N. C., 1927, pp. 33, 34a, 41a.

Fertilizer credit is likely to come at a higher rate. It may be secured through planters, merchants, or fertilizer dealers. In his study Eutsler found that 88 per cent of the Negro farmers from whom he secured records bought fertilizer on time. Their average fertilizer bill was $254, and the excess of time prices over cash prices was found to equal from 35 to 39 per cent interest, depending on the sources of the credit.[12]

There are two ways of looking at interest charges as high as 25 per cent on advances to cotton growers. They may be regarded as a species of merciless extortion by which the merchant may soon become wealthy. One writer[13] has called the credit and crop lien system "legalized robbery" and says that the borrowers, "cheated in the making of their contracts and in purchasing necessities . . . have been but the prey of sharks and harpies, bent upon keeping them in a state scarcely better than that of slavery." On the other hand, such high interest rates may be regarded as inherent in the speculative nature of cotton growing. It has been observed that many credit merchants fail while but few grow rich. R. B. Eutsler has offered an explanation of high interest rates in terms of the speculative risk involved:

Merchants ordinarily furnish fertilizer, equipment, seeds, and cash in addition to merchandise for consumption. Under such a system the merchant has assumed practically all of the risk for the success of the crops of the farmer to whom credit is extended. The only security available is a lien on the crop to be planted and grown during the same period in which the credit is granted. The risks which the merchant

[12] *Ibid.*, pp. 27–31.

[13] Scott, *Negro Migration During the War*, pp. 92–93.

assumes, then, include risk of production hazards of the crop, and market risks for the sale price of the crop when they are ready for marketing. In a one or two crop system of agriculture, the assumption of these risks is pure speculation. It is believed that this factor of risk is a major cause of the high cost of credit extended by time merchants.[14]

That the need of the grower of cotton and his lack of alternatives enable the supply merchant to charge higher credit prices must be admitted. On the other hand, if such business enterprises were uniformly successful, competition to secure high returns on investment would lead to a lowering of interest rates.

The Federal Farm Loan Act and the Joint Stock Land banks furnish only long-time credit. This form of credit on land enables an owner to improve his farm, or an industrious farmer to purchase land. It, however, makes no provisions for short-time credit needs of groups such as cotton growers.

DIVERSIFICATION

When one seeks to ascertain why growing cotton should require so large a percentage of credit, he is confronted with two types of answers. One answer explains the situation in terms of human defects—incompetence and lack of thrift and industry. Another answer may be found in the nature of cotton and the system under which it is produced. The cotton farmer receives his income all in a lump sum at the end of the season. Students of budgets will agree with agricultural economists that "if the year's income is all received at one time, it is difficult to make

[14] *Op. cit.*, p. 33.

the money last through the year, even when the same amount of money coming at convenient intervals might provide very well."[15] The dependence on the one crop also increases the chance of crop and market failures, and when the farmer is once thrown a year behind in his financing it is difficult for him to escape from a credit economy. Dairying or live stock farming provides an income at regular intervals, while a diversified system of farming also provides against total failure from any one crop.

Sufficient evidence exists to show that a large part of the credit used in financing cotton is borrowed for consumption rather than productive purposes. In a survey of 1,014 North Carolina farmers it was found that they used an average of $182.40 credit a year. Of this 50.9 per cent went for food, clothing, and home supplies, leaving slightly less than half for productive uses in the purchase of fertilizers, tools, and stock feed. For the landowners only 43.8 per cent of the credit secured was consumptive, while for the renters 62.4 per cent went for living expenses during the growing season. The effect of cotton culture is most significant to our discussion. It was found in a mountain county in which no cotton was grown that the average amount borrowed per family was $10 a year; in the cotton producing county each farm family borrowed on the average $436 a year.[16] "We feel rich after the crop is sold," one cotton tenant said, "rich, till we meet the people we owe."

The answer is, of course, the familiar one of lack of

---

[15] G. F. Warren, *Farm Management,* p. 108.

[16] Carl C. Taylor, and C. C. Zimmerman, *Economic and Social Conditions of North Carolina Farmers,* State Tenancy Commission, 1922, p. 31.

diversification. It is not only that cotton farmers have too little income distributed over the year from other crops, but that they have too many things to buy. "While we need and must have a better system of financing the cotton crop of the South," said Harvey Jordan, president of the Southern Cotton Association in 1906, "yet I tell you, the strongest financial institution for every farmer is a well-filled corncrib and smokehouse." Henry W. Grady, Seaman A. Knapp, Walter Hines Page, Booker T. Washington, all of the South's good friends who have been close to things of the soil, have talked and dreamed of the diversification of production on cotton farms. Yet no canon has been more persistently preached and more consistently breached than diversification. The old South had agricultural reformers and experimenters [17] who railed at the one-crop systems based on cotton and tobacco. In 1876 a cotton factor [18] wrote a pamphlet to his friends, the cotton growers, in which he plead for diversification in a strongly contemporary tone:

> We cannot break off from old ways all at once; but we are untrue to ourselves and our children if we do not make the trial, when our conditions show that the old ways are unprofitable. The course to our mind is plain: Plant only as much land in cotton as you can cultivate and manure thoroughly; and raise on your own place everything that is needed to feed your family and stock. Such luxuries for the table as must needs be bought, let them be provided for from your surplus butter, eggs, poultry, and dried fruit. You will then find that when your crop is sold you will have a

[17] See A. O. Craven, "The Agricultural Reformers of the Ante Bellum South," *American Historical Review*, XXXIII, 302–14.

[18] W. C. Grandy and Sons, Factors, *The Cotton Question Presented to the Cotton Trade*, p. 28.

balance on the right side of your factor's ledger, on which he will be glad to pay you interest until you are ready to use it in enhancing the value of your estate by improvements, or in giving a lift to a worthy son who has been made a man by the example of a father who knew how to make available what providence had vouchsafed him.

No triumph has been so confidently announced and so often delayed as the expected triumph of diversification in southern agriculture. The cotton crop lien system was thought to have been destroyed by the crop failure of 1868 which ruined so many furnishing merchants. Again in 1873 and in 1881 at the first Atlanta Exposition, speakers announced that the problem of southern agriculture was solved. Low prices from 1890 to 1894 led many merchants to refuse to grant further credit on cotton.[19] In every case the optimistic prophecies came to nought. In recent years the boll weevil invasion and the demand for foodstuffs imposed by the World War were thought by some to have changed southern agriculture beyond the possibility of relapse.[20]

In proof of the triumph of diversification, total figures on production are usually given. These, however, are not fitted to serve as an adequate index of diversification in that they tend to distribute the products of specialized fruit, truck, dairy, and poultry farms among the general averages and thus to make it appear that production on each farm has increased. Thus to say that South Carolina has exported $100,000 worth of poultry cannot be taken to mean that poultry in excess of normal consumption on South Carolina farms has been produced. It may

---

[19] Hammond, *The Cotton Industry*, pp. 157–58.

[20] See Chap. III.

as well mean that many of the farm families who do not produce these things are also unable to buy them.

A valuable index of the extent of specialized cotton farming was furnished by a special tabulation of the 1900 Census. It was found that at least 40 per cent of their income was derived from cotton by 19 per cent of the farms in the United States.[21] There were 1,071,545 of these specialized cotton farms in the South, forming 52 per cent of all the farms in the ten chief cotton states.

Another index of diversification is furnished by the percentage of farms growing specified crops. The year 1919 may be taken as a season in which southern farming was supposed to be well diversified, and the 1920 agricultural census may be analyzed in regard to farms growing certain products. In the ten chief cotton states there were 2,550,407 farms, with 92,645,980 acres of improved land, on which lived 13,367,407 persons. The average number of acres in crops on each farm was thirty-six; excluding Texas and Oklahoma it was twenty-seven. Seventy-three per cent of all farms grew cotton, producing 11,376,130 bales on 36 per cent of the crop land area in the ten states.

The percentage of farms growing specified crops is given in the following table:[22]

---

21 *Census,* 1900, V, Part I, liii. This is the latest agricultural census which classifies farms according to principal source of income.

22 This and the following table compiled from 1920 *Census,* V, 729–878. See *Foodless Farms,* E. J. Bodman. Leaflet.

TABLE VII

EXTENT TO WHICH FARMS ARE DIVERSIFIED IN TEN COTTON STATES

Rank		Number of Farms on Which Produced	Per Cent of Farms
	All farms	2,550,407	100
1	Corn	2,250,580	88
2	Eggs	2,049,996	80
3	Home gardens	1,953,160	77
4	Cotton	1,872,326	73
5	Dairy cow	1,605,074	67
6	Chickens	1,599,427	67
7	Butter	1,377,681	54
8	Hay and forage	1,167,489	46
9	Pigs	1,111,863	44
10	Sweet potatoes	1,069,110	42
11	Cane for sirup	607,962	24
12	Irish potatoes	545,014	21
13	Oats	364,901	14
14	Pure bred animals	134,441	5

A tabulation of the per cent of farms that grow cotton with the average sales per farm of poultry and dairy

TABLE VIII

COTTON, POULTRY, AND DAIRY PRODUCTS IN TEN COTTON STATES

	Per Cent of Farms in State That Grow Cotton	Average Sales Per Farm of		
		Poultry	Dairy Products	Total
South Carolina	93.4	$12	$12	$24
Mississippi	88.0	16	14	30
Georgia	87.0	13	18	31
Alabama	86.3	18	17	35
Texas	79.5	28	34	62
Louisiana	75.7	16	20	36
Arkansas	74.2	23	16	39
Oklahoma	57.4	50	62	112
North Carolina	56.7	25	22	47
Tennessee	30.3	58	34	92

products shows that the smaller the percentage of farms in the state growing cotton the higher the value of these products, except in Oklahoma and Texas.

When we come to account for the lack of diversification we are faced with a vicious circle. Lack of home supplies, raised on the farm, increases the need of consumption credit. On the other hand, the credit system calls for the production of cotton more or less to the exclusion of other crops. To begin stock raising, fruit farming, or dairying not only requires a larger initial outlay for capital, but the returns are regarded as less immediate and less certain. There exists the question of equipment needed in diversified farming. Fences and barns in the South are often too poor for the farmer to be able to care for live stock or crops of grain. For the small owner such outlays call for more credit. Landlords do not want to go to the expense of building good barns and fences, and the tenant cannot afford to make repairs that will revert to the landlord as soon as his tenure is up. Cotton culture, moreover, is comparatively simple in that poor methods will grow some kind of crop. There is a saying that any fool can grow cotton. On the other hand, the duties of management in dairy farming or stock raising are complex and require constant supervision. Whether equitable or not, a system of division of the product by shares between landlord and tenant has been worked out by custom and law for cotton and tobacco. No generally accepted system of share cropping has been worked out for more complex forms of farming.

Not only would it take a wise landlord to supervise a dairy or poultry farm, but he would start without precedent as to the system of tenancy. The question leads again to the human elements. The landlord and supply merchant are afraid to trust stock, poultry, grain, and feed crops to many of the renters. Grain may be eaten or fed to stock; pigs may take to the swamp or cane breaks; and blooded stock may die. But cotton is food for neither man nor beast. Some tenants may be inclined to steal, but the crop lien is a form of restrictive credit, limiting the sale of the cotton. Cotton is a bulky product difficult to cart away; it has to be delivered at a gin, and it is at the gin that a careful landlord checks up. For this reason the system has grown up in some localities of charging a cash rent, say $10 an acre, for corn while accepting a fourth share of the cotton. The effect of this form of rent has been to restrict the production of corn. It does not take the tenant long to realize first, that he must pay for his corn out of his cotton, and second, that the landlord is sharing his risks of production on cotton, but that he himself carries the whole load when he raises corn.

The crux of the matter may be summed up by saying that cotton made the one-crop system because it is the one cash crop. As W. J. Spillman has said, cotton has been the only product sufficiently high-priced to permit its shipment to distant markets. Oats and corn are too cheap, and wheat and hay meet with climatic difficulties. At any time at any place in the South, cotton is the one commodity that can be exchanged for cash—be it much or little. Every little town is a primary market, and cotton buyers from interior markets and cotton quotations from New Orleans and New York are ubiquitous. The cotton market is a world market—possibly the world's best organized market for any agricultural product. For one time in her history, when the New York Cotton Exchange closed at the outbreak of the World War, the

South was without cotton prices. The business stagnation and dismay of the section during those weeks can hardly be described.

And because cotton has its equivalent in cash, any time, any place, it is the mortgage crop. The larger the percentage of the farm planted to cotton, the more security has the lender of credit. Food furnished by the farm is of interest only to the actual farmer; the landlord and supply merchant receive their payments not in kind but in cash. Where diversified farming is acceptable to the landlord it may encroach upon the vested interests of the merchant and the furnisher. Why should the supply merchant encourage the growing of an abundance of food and feedstuffs at home when he has them for sale at the "General Merchandise" at a profit plus an interest charge? It is this factor which gives content to the term, "the cotton system."

There is an accumulated body of evidence to support the charge that the undiversified growing of cotton is determined by financial interests rather than by agricultural interests. The tenant farmer's wife of Alabama who wrote the *Atlanta Constitution*, pathetically inquiring how she and her husband or neighbors were to follow the universal advice to diversify when their landlord wanted only cotton grown, represents a type. A Georgia business man[23] wrote in 1914:

> If you want to get at the situation you must call a meeting of the owners of these rented farms and . . . the men who buy and sell what goes on and what goes away from the farm. These bankers, merchants, warehousers, fertilizer men,

---

[23] J. T. Holleman, *Is the South in the Grip of a Cotton Oligarchy?* pp. 9–10. Pamphlet.

mule dealers . . . find their occupations gone when cotton has no value. . . . Their one idea is to see that they [the tenants] produce every bale of cotton possible. If one of these owners lives too far away to look after renting and to attend to the gathering of the crop and the collection of the rent, some local man is made an agent for the purpose. He looks first to making . . . a good rent contract for the owner. . . . Then he sees to it that . . . the handling of all the cotton produced on that farm should come through his store or bank. . . . Fertilizers . . . mules . . . farming implements . . . all the supplies that go to the tenants on this farm are handled by him. . . . The farmer who raises his own supplies at home has no use for these men. [But] cotton farming can be done by any sort of poor white tenant and any sort of ignorant Negro. All these poor tenants need . . . is to have a little direction from town in the springtime, an arrangement by which all they eat and all their stock eats is furnished to them, and they can produce the cotton and carry it into town, where the lord of the manor will be ready to take it and sell it and pocket the lion's share and let the poor white tenants and Negro tenants have just enough to keep them alive until next year's crop. These men give no thought to the building up of the land but milk it from year to year of every ounce of cotton it will produce.

[As the result] today the southern farms are barren of the very things that should be their glory. You can get on a railroad train or into an automobile and ride a hundred miles without seeing a herd of cattle. When you do see cattle they are little tick infested creatures that no more resemble real cows than a tubercular cotton factory operative resembles an athlete. I have young men in my employ twenty-five years of age, born and raised in Georgia, who have never seen a mule colt. There is no grain, no hay, no poultry, no vegetable gardens, no orchards—except the peach orchards belonging to non-resident corporations—nothing that goes to make up a real farmer's home.

It is in years of cotton prosperity that the system flourishes most. When overproduction or crop failures leave notes, rent, and store bills unpaid, a cry for diversification goes up from southern business. But in the program for diversification there exists a fundamental conflict between the public needs of the region and the vested interests of those engaged in supplying the various forms of cotton credit. This has been caustically pointed out by a Texas lawyer:

There are those in every village, town and city in the South who constantly demand that cotton shall be planted in large acreage and have never at any time heretofore urged and are not now urging any decrease in acreage. They constitute that large and influential class engaged in local commercial pursuits. Their business is the farm supplying trade of the Southern States. They furnish the rations, either directly or indirectly, used by the farmers in producing crops, and they insist that those who eat the bread and meat and wear the clothes they sell on credit, produce a crop which can on any day, at some price, be sold for cash.

They may meet, yes, they do meet, in the Chambers of Commerce and proclaim that there must be a reduction of cotton acreage, and from those places they proceed directly to the cuddyholes in their places of business and write chattel mortgages for the poor devils to sign covering the cotton crops from one to five years and in addition, everything from the pig to products raised by the children. Listen to the speeches in the Chambers of Commerce, and in the banquet halls where a few most carefully selected farmers are entertained, and go directly from there to the mortgage records and see the work that is really being done. Always, everywhere, in the private counting houses, the demand is made that cotton must be produced so that the debts can be paid.

Not one word is said in private conference with the farmers about reduction of cotton acreage.[24]

It may be that the vested interest in cotton culture is more apparent than real. If the change were made to a well-rounded system of diversified agriculture with cotton as a cash crop, the supply stores and purveyors of credit would likely find a place in the economic organization equally remunerative because less hazardous. But, like the introduction of machinery, the adoption of a new agricultural economy would cause a shock resulting in the maladjustment of many economic units. Since changes in industrial organization are always engineered by those in possession of the credit facilities, to expect croppers, tenants, or even small owners to carry out an adequate program of diversification is to reproach them for not being able to lift themselves by their own bootstraps. The change when it comes will have to be engineered from above by bankers, landlords, and supply merchants. A general change in the cotton system would involve a social crisis such as a continued depression in the cotton market, or a withdrawal of cheap tenant labor as has occurred in some areas since the Negro migrations. Such an adjustment would result from continued loss to supply merchants and landlords; it would be painful and of long duration.

Many southern thinkers are coming to feel that the break-up of the plantations into small owned farms offers the only chance for diversification. The plantation with its organized supervision is the more efficient producer of cotton, but it does not provide for the home living from

[24] T. N. Jones of Tyler, Texas, in an open letter to *Dallas News*, Feb. 17, 1928.

the farm for its workers. The least changes toward diversification may be expected where the plantation or absentee landlords prevail. Many of the landlords who have been successful farmers "now reside in the towns, expecting their tenants to do just as well as they did but at the same time demanding more and more money from the old place."[25] This of course means a concentration on cotton and tobacco. A Mississippi Flood Director for 1927 writes of conditions in the Delta:

> Portions of this region have been given almost entirely to the raising of cotton. No richer land can be found anywhere than in the Mississippi Delta. Areas that planted 100 per cent cotton are mortgaged to such an extent that it is a question as to whether more than a very few will be able to survive the late overflow. It was necessary for the Red Cross to feed their tenants for almost twelve months, and at this time many of them are experiencing much difficulty in obtaining credit. There are few planters in the Delta who raise their own feedstuff for their hogs and cattle sufficient to supply their farms. Invariably when you strike such a farmer he has money in the bank and seldom do you find such a farmer with a mortgage on his farm.
>
> I have seen many farmers in the hill district go out and clear land, and establish a new farm, and apparently they seem to be making money for the first two or three or four years. While the land is fresh it would yield three-quarters of a bale to a bale per acre. Their tenants are able to pay out, but after five, six, seven, and eight years the lands begin to fail. It would take three to four acres to make a bale, and after having planted their farms constantly in cotton, most of the landlords lost their farms, had to give them

[25] Charles E. Gibbons, "Farm Children in Oklahoma," *Child Labor Bulletin,* May 1918, p. 40.

up, and the supply merchants had to take charge of them. Many of the farms have been thrown out, and the landlords are either somewhere in public work or renting themselves.[26]

The greatest changes made in the cotton system have occurred in areas inhabited by white and black land owners, intermingled, who have been aided by bankers. Live stock, dairy cows, poultry have been purchased; cover crops such as velvet beans, soy beans, cow peas and lespedeza hay have been grown, and the acreage of cotton has been reduced. By rotating so as to plant the cotton each year on soil previously occupied by legumes, it has been possible to increase the yield of cotton while reducing its acreage. Studies by the Department of Agriculture have shown that such intensive culture increases the cost per acre of producing cotton but materially reduces the cost per pound. The forcing methods of cultivation are also the ones that have proved most effective against the boll weevil. To such an extent is this true, a president of a fertilizer company with fertilization and the weevil on his mind said that successful cotton growing is tending to change from an extensive field type of cultivation to an intensive garden type with a pharmaceutical department attached.

MARKETING

The methods by which cotton is marketed[27] are inher-

---

[26] Letter from L. O. Crosby, Picayune, Miss., March 12, 1928.

[27] Dealing in cotton constitutes a definitely recognized profession in the South, employing a number of factors, brokers, dealers, and cotton buyers. Around this trade surrounded by the hazards of the market has grown up a set of attitudes which may be called a speculative complex. The concern of the present discussion is with the relation of the grower to the marketing of his product.

ent in the credit system. The crop lien, as said above, is a form of restrictive credit which assigns to the creditor the control of the sale of the product. Accordingly, cotton produced on rented land or on credit arrangements is sold through the landlord or merchant. Often in the Cotton Belt ginners and supply merchants are cotton buyers, having connections with the larger cotton companies. The demands of the creditors and the fact that cotton is practically the only cash crop for many farmers, tend to throw the product on the market as soon as it is ginned. An observer of southern agricultural conditions thus sees the interaction of producers with buyers in the primary markets:

> I saw the farmers coming into the markets this fall by hundreds with the products over which they had sweated and toiled all the year; a great unorganized procession . . . without a common purpose and no group direction. . . .
>
> Once in market towns they find themselves the one unorganized group in a world more or less highly organized. [The cotton farmer] finds himself everywhere under pressure and in a hostile atmosphere. . . . Each man must deal separately with this powerful and intricate machine for buying and financing his crops. He finds that he is competing in the market with his neighbor farmer, . . . thus lowering the price for both. . . . Often he cannot wait; the merchant and the banker are clamoring for their money, . . . the result . . . prices are often forced below the cost of production.[28]

Since this study was written, A. B. Cox's *The Marketing of Cotton* and R. H. Montgomery's *The Coöperative Pattern in Cotton*, a history of the Texas Cotton Coöperative, have been published. Both are of great value to the student of cotton culture.

[28] Ray Stannard Baker, *New York Evening Post*, March 10, 1921.

The rush of cotton to market [29] can be seen in the fact that 80 per cent of the cotton handled during the whole year is marketed from September to February. Of the 11,300,000 bales marketed during the whole year of 1921, over 8,000,000 were disposed of from September to February. In 1922 over 10,000,000 out of 12,443,000 bales were marketed during the same period. In 1926, 10,250,000 bales were thrown upon the market during October, November, and December, on the heels of a 2,698,000-bale shipment in August and September. The market sagged under the weight, and cotton brought lower prices in December. As a result, the facilities of the Cotton Belt, poor at best for handling the crop, suffer a near breakdown. Defective terminals, inadequate storage, and a general indifference result in exposure that to any other commodity would spell ruin.[30] It is not possible to state with accuracy the losses to producers from exposures, but it is known that millions of pounds of worthless staple are scrapped from bales, caked and rotted from mud and water. In the season of 1918–19 the loss from bales sent to the pickeries from the New Orleans market amounted to 1,142,000 pounds, equivalent to

---

[29] " . . . the cotton crop . . . has been marketed with a degree of foresight and regularity that can best be likened to a stampede of Texas steers.

" . . . The orderly marketing of the cotton crop can never be accomplished by a heterogeneous, motley-minded mass of farmers. . . Anyone attempting to bring order into this chaotic state as it exists today, has truly a Herculean task before him." D. B. Osborne, *Manufacturers' Record,* Dec. 7, 1922.

[30] Statement of former Governor John M. Parker of Louisiana, President of New Orleans Cotton Exchange, before National Agricultural Conference, at Washington, D. C., Jan. 25, 1924. Mimeographed Press Release.

nearly a pound and a half for each bale of the press receipts at New Orleans.[31]

If it were left to be determined by the nature of cotton itself rather than by the financial weaknesses of the growers, the marketing of cotton might well be evenly distributed throughout the year. Cotton, as a matter of fact, keeps better in storage than any other farm crop. "Nearly all banks," writes Gilbert H. Collings, "are willing to advance money on cotton at liberal terms if the cotton is properly stored and insured. In fact they prefer it to real estate and will accept it as collateral as readily as government bonds."[32] Most warehouses, however, belong to mills, middlemen, factors, and the non-farmer group. One-half the warehouses in Texas, for instance, are located in two cities, Galveston and Houston. Georgia has the best distributed system of warehouses, but in most areas warehouses offer comparatively poor service for high charges to farmers because of the small volume of business.[33]

There exists justification for Carl Williams'[34] statement that "cotton has been and still largely is a city man's crop" in the fact that cotton classing is a technical art of which the farmer is ignorant. Since cotton cannot be used on the farm or by retail consumer, there exists no possibility of the growers' dealing with individual consumers. The modern cotton trade, then, is organized on the basis of assembling a large quantity of fibre of certain qualities to supply vast manufacturing establishments. There exists, thus, for the cotton pro-

[31] Report of New Orleans Cotton Exchange, 1919.

[32] *The Production of Cotton*, p. 155. [33] *Ibid.*, pp. 156–57.

[34] President American Cotton Growers' Exchange, before Agricultural Conference, 1924. Press Release.

ducer only a wholesale market manned by technical experts in the art of classing cotton. The lack of what may be called retail competition, the fact that the farmer has no alternative but to sell to the manufacturer's representative, the cotton buyer, whose methods of classing he may not understand, is responsible for the farmer's suspicion of cotton exchanges and his demand for special legislation in regard to marketing.[35]

That there exist foundations for suspicions of the cotton marketing system is pretty generally known throughout the Cotton Belt. A study of cotton market conditions in small towns in Oklahoma[36] found that in the same market on the same day, the same grade of cotton brought widely varying prices, depending on personal factors. This situation may be owing to the ignorance of the county buyer in the classing of cotton. He pays the grower of poor fibre too much, and he pays the grower of long staple too little. It has been shown that the general merchant pays a higher rate for credit than non-credit cotton, thus offsetting to a slight degree high interest charges.[37] It is Mr. Coker's opinion that:

> The county buyer pays the same price for all cotton that is brought to him. This is the fault of the large cotton houses, local buyers, and cotton mills themselves. With a few notable exceptions these have taken little or no interest in the production of superior cotton, though they could almost immediately increase the length and improve the char-

[35] See O. F. Cook, *Relation of Cotton Buying to Cotton Growing* Dept. of Agriculture Bulletin 60.

[36] See *Studies of Primary Cotton Market Conditions in Oklahoma,* Dept. of Agriculture Bulletin 36.

[37] C. J. McConnell, *A study of Cotton Market Conditions in North Carolina,* Dept. of Agriculture Bulletin 476, p. 17.

acter and grade of the entire crop by seeing that each individual grower was properly compensated for superior quality in his product.[38]

On the other hand, it is considered legitimate in many instances to underbuy cotton. "The failure to discriminate in price to the farmer is so general that many buyers do not consider it dishonest, but look upon it merely as one of the ways of increasing the profits of their business." [39] The failure to pay the market price for the staple is due, it has been suggested, to the restriction of competition. "In some towns . . . in the Cotton Belt there is a general understanding among these men [buyers] whereby each enjoys the exclusive right to do business with certain farmers. . . . For instance when a farmer brings a load of cotton to town, one of them will walk out and place his foot upon the hub of the wagon and begin talking to the farmer. This is the accepted sign that the farmer is his customer." [40]

It was shown in the North Carolina study that the producer who knew the class of his cotton obtained on the average $1.15 per bale more than the farmer who was not furnished such information.[41] If this can be accepted as generally true the growers must lose from twelve to sixteen millions of dollars a year because of ignorance of grade. Carl Williams is authority for the statement that the Mississippi and Arizona coöperatives "have consistently sold cotton at from $10 to $15 a bale above country market prices. The Oklahoma and Texas

---

38 Letter of David R. Coker, Dec. 29, 1926. In the files of the Rural Social Economics Library, Univ. of N. C.

39 Dept. of Agriculture Bulletin 60, p. 7.

40 Charles E. Gibbons, *op. cit.*, pp. 43–44.

41 Dept. of Agriculture Bulletin 476, p. 18.

short-staple associations whose cotton is of lower value have developed a differential of from $5 to $10 a bale." [42]

The Cotton Coöperatives are the attempts of the growers to break through the cotton system. Like many social changes, the organization of coöperatives grew out of a crisis—the fall in the price of cotton after the war. The first to be organized was the Oklahoma Association in April, 1921. The movement grew until in 1925 there were fifteen Cotton Coöperatives in thirteen states with a total membership of 284,867. They have formed a federation in an overhead body, the American Cotton Growers' Exchange, which coördinates the activities of the various associations. Within four years from the founding of the organizations they had handled 3,140,928 bales of cotton.[43] Conflict with the established cotton system has been evident throughout their history. Men who have signed up their cotton have not been able to deliver it to the Coöperatives because of credit arrangements and crop liens. Many more have been kept from joining by this factor. The tenant has no option; his fate is in his landlord's hands. If the landlord joins the Coöperative, the tenant, in debt to the supply merchant, cannot wait for any part of the payment for his share of the crop. In addition to the individualism of the farmer, the Cooperatives have had to fight the covert, and often open attacks of the vested interests in the cotton trade. Supply merchants and cotton buyers have been hostile toward these efforts, on which depends the future of the cotton system in the South.

---

[42] Statement before National Agricultural Conference. Press Release.

[43] George O. Gattin, *Coöperative Marketing of Cotton,* Dept. of Agriculture Bulletin 1392, pp. 1–2, 17, 25.

FIELD LABOR OF WOMEN AND CHILDREN

Another factor so far inherent in cotton culture is the large amount of labor required. Not only is a large amount required but it is cheap labor. Cotton is a cheap labor crop, as has been suggested, largely because it cannot be grown by machinery. Machinery does the work of many men and pays their labor returns to its owner. Machine agriculture enables the farmer to cultivate more acres and thus add to his increased labor returns, increased returns on capital. The amount of cotton a farmer can grow is limited by the amount he can pick. Cotton has so far defied machine harvesting. This defiance limits the development to any great extent of mechanized methods of planting and cultivating. It is useless to plant more cotton than can be picked.

The greater amount of labor required on cotton farms has been shown by comparative farm management studies.[44] A group of cotton farms in Sumter County, Georgia, averaging 73 acres in crops were found to require on the average 38 months of man labor per year. Dairy farms in Dane County, Wisconsin, averaging 81 acres required 22 months of man labor, and grain and live stock farms of 93 acres in Clinton County, Indiana, required only 19 months per year. A cotton farm 85 per cent as large as a live stock farm thus requires twice as much man labor. The labor services required are also more intensive in that they are distributed in peak loads. As a result the average cotton farms have the smallest acreage of improved land of any farms except those

[44] H. W. Hawthorne, et al., *Farm Organization and Farm Management in Sumter County, Georgia,* Dept. of Agriculture Bulletin 1034, pp. 23 ff.

devoted to truck growing. The average for the United States is 78 acres of improved land per farm, while all the cotton states except Texas and Oklahoma have less than 45 improved acres.[45] If large farms make possible economies, cut down overhead, and pile up incomes, the cotton grower farms too meanly on too small a scale.

Another reason often given for cheap labor in cotton is that the cotton grower is inferior in that he is willing to accept a low standard of living. Although efficient, it is contended, they are numerous enough to overproduce cotton. After they have depressed the price of cotton below a decent standard of living they continue producing it, since it is impossible to depress their standard of living low enough to cause their withdrawal. The form this argument takes is interracial competition with Negroes on the plantation. The view has been well expressed by a cotton farmer himself writing from Georgia in 1914:

> One great reason, why cotton is so plentiful and cheap, is that on the great plantations of middle Georgia, middle Alabama, Mississippi and others of the richest regions of the South, is it is grown by Negroes who get for their labor, only enough to maintain a bare, brute subsistence. . . .
>
> They are also intimidated by the sentiment which is kept alive among you against them, and which costs you and your children more than it does them. The few, rich non-resident plantation aristocrats who get the benefit of this cheap, this mule cheap labor, against which you are putting your wives and children in the fields, are not, and never would be, able to intimidate these Negroes by themselves. The Negroes work year after year making cotton to make yours cheaper, and never daring to ask for a settlement in many cases because you make them afraid. Let a voice go up from every cotton

[45] Dept. of Agriculture *Yearbook,* 1921, p. 491.

field in the South, that white women and children are no longer to work against Negroes who are reduced to the mule level for the benefit of non-resident plantation owners living in New Orleans . . . Europe, and everywhere except where they have to hold a plow handle.[46]

Family-size farms are "such as require from 12 to 24 months of man labor" over the period of a year.[47] If this labor is relatively unskilled, of a manual rather than machine tending type, and if the work comes in seasonal peak loads, the farm will come to be tended to a greater extent by women and children. These conditions, as we have seen, are met in the culture of cotton. The economic status of the family, the available supply of wage labor, and the prevailing cultural attitudes are factors affecting field work for women and children. While more tenants than home owners, and more Negro than white farmers work their women in the fields, the practice is more a factor of income and standard of living than of home ownership or race. A reporter to the 1880 Census,[48] from Conway County, Arkansas, "where four-fifths of the cotton is produced by white people chiefly on small farms," laid the increase in cotton culture in that region to the fact that "it is a crop in which all the children and women of the family can earn their own support at home without undue exposure or hard labor."

In the 1920 Census of Occupations, 1,084,128 women in the United States were listed as engaged in agriculture and kindred pursuits. Of these, 869,416, or 80 per cent, were found in the ten chief cotton growing states. The large majority were listed as working on the home farm,

[46] C. D. Rivers, *The Empire of Cotton.* Pamphlet.

[47] H. W. Hawthorne, et al., *op. cit.,* p. 24.

[48] "Cotton Production," V, 618.

while the next largest group was composed of wage hands.[49] That the number of women farm workers is seriously underestimated is suggested by the nature of the questions asked as to occupation. With a smaller farm population in 1910 there were found 1,807,506 females over ten years of age engaged in "agriculture, forestry and animal husbandry," 83 per cent of which were in the ten chief cotton states.[50] According to the 1920 figures, however, 19.8 per cent of all females on the farm over ten years of age in the ten cotton states served as field laborers. Of every hundred women field laborers 68 were Negroes and 32 white in 1920. In 1910, 60 were Negroes and 40 whites.

In a report of family labor employed per farm October 1, 1927, the South Atlantic states averaged 3.85 and the South Central 3.62 persons as compared with 1.79 for the Western states, 1.82 for the North Atlantic and North Central states, and 2.51 for the United States as a whole.[50a]

The distribution of the field labor of women according to states is shown in the following table. It will be seen that the percentage of women working in fields tends to vary with the percentage of farm lands devoted to cotton. Texas and Oklahoma, because of the large scale farming and comparative scarcity of Negro farmers, offer the two exceptions.

---

[49] "Occupations," *Census* 1920, Vol. IV, chap. VII.

[50] "Owing to changes in the date of enumeration from April to January and to changes in the wording of the instructions, the number of persons, especially women and children, reported as farm laborers was much less in 1920 than would have been the case had the 1910 method been followed." Note: *Statistical Abstract of the United States*, 1925, p. 46.

[50a] *Crops and Markets*, V (No. 10, Oct., 1928), 362. This refers to farms of crop reporters.

TABLE IX

PERCENTAGE OF WOMEN LISTED AS AGRICULTURAL LABORERS IN 1920 IN RELATION TO COTTON ACREAGE IN TEN CHIEF COTTON STATES

	State	Women on Farm 10 Years Old and Over	Women Listed as Agricultural Laborers	Per Cent	Per Cent of Farms Growing Cotton	Per Cent of Improved Acreage in Cotton
1	South Carolina	369,140	126,991	34	93	43
2	Mississippi	459,094	132,639	29	88	31
3	Alabama	473,044	122,755	26	86	26
4	Georgia	588,287	129,027	23	87	36
5	Louisiana	276,163	52,611	19	75	24
6	Arkansas	398,731	66,310	16	74	28
7	North Carolina	528,227	80,472	15	56	17
8	Texas	786,398	99,978	12	79	37
9	Tennessee	463,325	36,336	08	30	6
10	Oklahoma	337,284	22,161	06	57	15

That the agricultural labor of women and children is much more prevalent in specialized cotton areas has been shown by investigations. In selected localities studied in North Carolina, children worked in the field in 65 per cent of 219 families, and in 75 per cent of 270 Negro families.[51] In a Texas survey [52] out of 1,561 children from six to sixteen, 75 per cent were reported as doing field work. One-third of these children were ten years of age and under, and 57 per cent were twelve years and under. They averaged ten hours a day picking cotton. More than half of the white mothers and 85 per cent of the black mothers in each Texas county studied worked in the field.[53] "To some extent," says a Children's Bureau report, "the amount of rest a mother can have before and after confinement is determined by the time of year or by the stage of the cotton crop upon which depends the livelihood of the family." [54]

It may be said that cotton culture, the tenant system, credit, and the crop lien have resulted in a standard of living for southern farmers of which the field work of their women and children is an index. One southern journalist has said without any intent to be a phrase-maker that cotton is the by-product of large families.

The questions of the living from the farm implied in diversification, the percentage of the family income spent for credit, and the extent of labor of women and children lead logically to a consideration of the cotton growers' standard of living.

---

[51] *Rural Children in Selected Counties of North Carolina*, U. S. Dept. of Labor, Children's Bureau Publication 33, p. 49.

[52] *Welfare of Children in Cotton-Growing Areas of Texas*, U. S. Dept. of Labor, Children's Bureau Publication 134, pp. 23–5.

[53] *Ibid.*, p. 44.

[54] *Rural Children in Selected Counties of North Carolina*, p. 34.

CHAPTER VIII

HOW THE COTTON FARMER LIVES

OPINIONS

THE CONFLICTING attitudes as to the standard of living he has, or should have, might lead one to believe that the cotton farmer is the victim of a preconception of his rôle. When he arrived on the scene after the Civil War, cotton had merely changed its status from a slave crop to a cheap crop. Cotton labor was definitely classified as cheap labor. From Phillips' accounts of Negro slavery and from contemporary records of the small upland cotton producers variously called "poor whites," "sand hillers," and "clay eaters" one gathers that their standards of living must have been low indeed. Edward Atkinson, the New England spinner, by 1861 had anonymously published a pamphlet speculating on the possibility of *Cheap Cotton by Free Labor* after the war. Atkinson arrived at $20 per month as the estimated upkeep of a Negro slave and his family and noted that with cotton at ten cents the owner had continued to buy more land and more slaves. "It is therefore evident," he concluded, "that the owner of land can afford to pay $20 per month wages, and that there is a class of poor white laborers (poor white trash comprising the large majority of the cotton states) at hand to whom such pay would be an income never dreamed of." [1] He did not hold

[1] Pp. 11, 3–4.

to the view accepted by some that it is dangerous to the health of white people to work in the cotton field, and cited Olmsted's *A Journey in the Back Country* to the effect that:

> The necessary labor and causes of vital fatigue and vital exhaustion attending any part or all of the processes of cotton culture does not compare with that of our July harvesting; it is not greater than attends the cultivation of Indian corn in the usual New England method. I have seen a weakly white woman the worse for her labor in the cotton field but never a white man; and I have seen hundreds of them at work in cotton fields under the most unfavorable circumstances, miserable, dispirited wretches and of weak muscles, subsisting mainly, as they do, upon corn bread. Mr. De Bow estimates one hundred thousand white men now engaged in the cultivation of cotton, being one-ninth of the whole cotton force of the country.[2]

The opposing attitudes and opinions as to the standard of living of the cotton farmer are plainly motivated by the conflict of interest between the spinner, desiring cheap cotton to manufacture, and the producer, convinced of inadequate returns. A clash[3] that occurred in a hearing on cotton crop estimates in 1905 between Representative Lovering of Massachusetts, a cotton spinner, and Representatives Clayton and Burleson, both southern planters, is significant:

> Lovering: . . . "Here is a crop [cotton] paying the planter 100 to 150 per cent . . ."

---

[2] *Ibid.*, cited p. 10.

[3] *Hearings of the House of Representatives, No. 45, Session 1905–6*, Dec. 19, 1905, p. 15.

Clayton (interrupting): "I challenge that statement. I know something about it and I know it is not correct."

Burleson: "If it pays so large a per cent to the farmer and so small a per cent to the spinner, how do you account for the fact that the planter of cotton is almost invariably a poor man, while the spinner is almost uniformly a wealthy man."

Lovering: "I should dispute that with you. With 625 millions going into the South for its cotton crop, who gets it?"

Clayton: "The gentleman's statement as to the profits derived by the southern planters from raising cotton is so absurd that it would make the ordinary plantation mule down in Alabama or Texas laugh, and the mule is a solemn animal and does not generally laugh.

"I was born on a cotton plantation and raised on one. I know something of the hardships and disasters that come to the cotton grower, the risks that he takes, the hardships he endures, and the many disasters that often come to his crop.

" . . . the cotton plant is the most tender field plant that grows. It is susceptible to more disasters and requires more of human manual labor to produce it than anything else. . . . You cannot make cotton with machinery as you can wheat and corn and other crops."

This clash of interests is more acute between the American growers and the English spinners. Southern merchants and bankers have come to feel that whenever cotton goes high enough to furnish an adequate standard of living for its producers, the English manufacturers begin to talk about cotton famine and to encourage the growing of cotton in the colonies. A writer in the *Manufacturers' Record*, organ of southern finance, expresses these views:.

There is probably no other crop in the world against which there has been such a tremendous fight, especially on the part of foreign buyers, for the purpose of breaking down prices and holding cotton growers to starvation wages. For a hundred years the English cotton manufacturers especially, and the English Government as well, have put forth their utmost power to break the price of cotton in America and to produce cotton in other parts of the world in order to lessen their dependence upon the South.[4]

England's fear of expensive raw cotton and her hope of escape are both based on the low standards of living of the more backward people. A great part of her textiles have been cheap cotton cloth exported to China and India. When the price of American upland middling reaches certain heights, not only the profits but the exports themselves are in danger of being cut off. England has sought to avoid this difficulty by attempting to stimulate cotton growing by the Empire's supply of cheap labor. The British Cotton Growing Association, founded in 1902, has been granted subsidies by the Imperial Government up until 1916. "The Association began operations in West Africa by engaging a number of experts from the United States. . . . Large areas were acquired and put under American cotton on the plantation principle with native labor, but it was eventually found that this method was not likely to succeed under West African conditions; and therefore the policy adopted was to encourage cotton growing as an industry . . . conducted by the native on his own land."[5] This area was later abandoned; cotton culture was introduced into other cheap labor areas, and by 1924 the new fields un-

[4] "Fifty Years of Southern Progress," Dec. 11, 1924, p. 336.

[5] W. H. Johnson, *Cotton and Its Production*, p. x.

der the protection of the Association produced 261,900 bales of cotton.[6]

One gathers, also, from the writings of such an English authority as Mr. John A. Todd that the world's cotton supply must come from agricultural labor with low standards of living:

> Cotton has always been regarded as a cheap-laborer crop, that is to say, a crop that can only be profitably cultivated where there is an ample supply of cheap laborers. Such a supply of laborers was obtained in the United States by the introduction of slaves, who, though neither very industrious nor efficient, could be trained to the necessary processes of cultivation and picking. Indeed it is admitted that a good Negro is the best cotton cultivator, if he can be persuaded to do his best. But since the liberation of the slaves, good Negroes have become almost the exception; the average "Nigger" has an incurable aversion to steady and especially to prolonged labor. . . . The scarcity of labor has only resulted in raising the general level of wages, and enabling the Negro to adopt a higher standard of living, and copy the luxuries and vices of the white man. . . . The contrast between all this and the position of the Egyptian fellah, with his unlimited capacity for patient plodding work from morning till night, for almost seven days a week, and from one year's end to another, on a wage of less than a quarter of that of the American Negro, which yet enables him to maintain a standard of living that makes him the healthiest and strongest agricultural laborer in the world, is painful in the extreme.[7]

This view regards the human factors as merely a means to the production of cotton.

Mr. Todd[8] reaches the conclusion "that in many cases

[6] *Ibid.*, p. xi. [7] *The World's Cotton Crops*, pp. 107–8.
[8] *Ibid.*, p. 113.

at least, the profits of cotton growing are by no means so large as we were apt to imagine. The position is practically this, that under such conditions cotton cannot be grown at a profit *if all the labor it requires has to be paid for.*[9] Unless the small planter has a large family to do part of the work, cotton at present prices will not pay. Such conditions may be all very well for the Negro, whose standard of living (for his children) has always been low; education, for example, is only now beginning to be thought necessary; but they will not do at all for the small white planters, who, in Texas especially, are very numerous, and who might have proved the hope of the cotton trade under more favorable conditions. In other words, it means that cotton is, and must remain a 'black man's crop,' not a white man's."

The conflict between the views of spinners and producers of cotton has approached nearest bitterness on this very phase of the field labor of women and children. In an address delivered in Brazil in 1921, Arno S. Pearse of Manchester, England, general secretary of the International Federation of Master Cotton Spinners and Manufacturers' Associations, said in speaking of the cotton outlook:

> . . . We were told on the authority of the president of the American Cotton Association that the American cotton farmer would no more continue to allow his wife and children to work in the fields. With such fantastic ideas it cannot be expected that there will be an increase in the cotton acreage of the United States of America.[10]

---

[9] My italics.

[10] Statements cited in *Manufacturers' Record* for July 31, 1924.

The phrase "fantastic ideas" was widely quoted in the South and brought forth a storm of rebuke.

A report of C. P. Ellis and Company,[11] cotton buyers of New Orleans, Louisiana, issued in the summer of 1919, commented on the high cost of cotton production shown by government surveys as follows: "It is well known that the great bulk of the cotton crop is raised by the small farmer, whose wife and children constitute his only help; hence, all these ridiculous estimates of cost of production are unworthy of consideration."

The lack of economic value implied to the help of the farmer's family, together with the brusqueness of the statement, caused it to meet with strong editorial protest. One paper spoke of the "enemies within." In this connection President Bradford Knapp of Oklahoma A. and M. College, made the following comment: "I know the world wants cheap cotton to clothe its nakedness, but may God forgive the man who wants it at the price of the labor of women and children in the field."[12]

Those closer to the culture are often found prone to indignation over the living conditions of the human factors in cotton. A North Carolina pamphleteer in the 1880's views the standard of living in terms of population renewals:

> Any person starting from Nag's Head—the extreme East and traversing the state to Paint Rock, in the West would be constrained at every step as he viewed the farms around him to exclaim: "My God, how do these people live!"
>
> Nature intended every man to be the father of four or five children, at least; but how is he to support them, at the present price of provisions in North Carolina? Cotton

[11] *Loc. cit.*

[12] *Loc. cit.*

puts him in one of two very unfortunate predicaments. He must either violate one of the fundamental rules of natural law and thereby incur the wrath of God, deprive himself of happiness, and his state of citizens, or he must observe that rule at the risk of imposing misery upon his offspring, and pauperism, and hence crime upon his community; yes, in thousands of instances cotton has put man in this dilemma.[13]

T. M. Young, an English traveler who visited the South in the early 1900's, has left his impressions of the housing and clothing of cotton farmers in North Carolina:

Dotted about at wide intervals are the wooden cabins of the peasantry. Some of these tiny dwellings are whitewashed, but most of them have never known either paint or whitewash, and never will know them. Very poor and mean-looking they are, but the blaze of roses which you may often see beside the doors and the space and purity around them redeem them from the appearance of squalor. It is in homes and amid surroundings such as these that the population has been bred from which the southern mills are drawing their labor, and the people have that fine physique which one finds in Irishmen bred in even the poorest country cabins.

In Piedmont sections [I was told that] farmers brought up large families for almost nothing; the cost of clothing where men work in a cotton shirt and a pair of cotton trousers for nine months in the year, was very much less than in the North; and fuel when it was needed, could be had for nothing in the nearest wood. Many of these people hardly ever saw money before the mill started, and now according to my informant, they hardly know how to spend it. At this mill they were paid in cash every week end. Much of what they earned was "wasted on tawdry finery."[14]

[13] W. R. Henry, *Cotton and the Commission Merchants*, p. 12.

[14] *The American Cotton Industry*, p. 15.

Daniel J. Sully of Providence, Rhode Island, was a cotton speculator who cornered the market in 1903–4 and pushed cotton to the high level of 17 cents. He had made several trips through the South before he failed and expressed great interest in the conditions of the cotton farmer. Writing in a popular magazine in 1909 he assigns the lowest standard of living in America to the cotton grower:

> American cotton planters, proprietors of the greatest gold-producing staple in the world are poor. They are in practical servitude . . . themselves absolutely subservient and the poorest paid toilers in the United States.
>
> Our greatest asset is our greatest humiliation. Cotton is king, but it is a badly served monarch. . . . It does not enrich but rather impoverishes the Southland. An enormous profit is made somewhere in the progress of the cotton to consumer. Every year cotton goods to the value of nearly six billion dollars are turned out from the 125,000,000 spindles in the world. But the poor farmer in the cotton fields sees but a pitiful part of the multiplying fortunes attending the migration of cotton goods around the earth.
>
> The ordinary grower of cotton cultivates twenty acres, producing one-half a bale to the acre. Unfortunately in too great a majority of cases he is a tenant farmer. Of his ten bales, the result of his year's toil, five must go to the owner of the land. The working farmer for his product gets, we will say, ten cents a pound or fifty dollars a bale; his twelve months of effort and expense bringing him in a gross revenue of $250. This is an insignificant total for the men who among others produce the commodity that controls the world.
>
> Out of that $250 he must provide for his family, himself, and his mule, and make provision for the ensuing times of planting and cultivating. Fully sixty-five per cent of American cotton crop is produced by this struggling method.
>
> The whole world is combined against the southern farm-

ers to keep down the price of their product. . . . The mill owner and the weaver want cheap cotton. Tenant farmers are under an enormous handicap and the landowner knows that high prices mean ability on the part of the tenant to get out from under the yoke and become independent; this the landowner does not wish. The southern farmers must combine; must restrict their crops along the lines of actual supply and demand; must abandon their slipshod methods of farming and resort to scientific processes; must diversify their crops and stand together.[15]

Southern opinion has often pointed to the burden the cotton producers' low standards have placed on the whole section. David R. Coker, outstanding planter of Hartsville, South Carolina, has said in an interview:

"The South hasn't had enough pay since the Civil War to advance a single step in civilization. Whole families lived on 25 cents a day in the years of five cent cotton. About 2,000,000 farm families produced the 1922 crop of 10,000,000 bales. That means five bales a family, worth $600. The other farm products are worth perhaps $300, making the total family income $900. The case of the cotton tenant is still worse. His total family income in South Carolina in 1922 was $365 or a dollar a day.

"Perhaps we haven't all realized how wretched an existence the small cotton grower has been forced to lead. Since the Civil War, whether white or colored, he has been the equivalent of slave labor."

Whenever and wherever I travel [adds the writer], through the Cotton Belt east of the Mississippi the things that haunt me most are the ragged tenants, wrinkled wives, half fed, anæmic children, and the wretched hovels in which

---

15 "King Cotton's Impoverished Retinue," *Cosmopolitan,* Feb. 1909. pp. 253, 258, 261.

they live, whether white or Negro . . . for the past fifty years cotton has been produced out of the very life blood of the South.[16]

A Georgia farm owner gives the attitude of the cotton producer on living standards and child labor:

I want to make some money and have tried, but cannot do so growing cotton. I am sore on cotton farming, as it is impossible to be a thrifty citizen and grow cotton. There is no wage in it. You cannot grow cotton to profit by hiring your labor, and where a man does not work his children from six years of age up in the field, he cannot come out of debt and also have the necessities of life. He could not do it if his life depended upon it.

There is only one way for a planter or farmer to make money with the use of hired labor, and that is to have a grab or commissary and keep books—always careful that no laborer exceeds his account, and making sure that at the end of the year he has gotten it all, and his labor has "just lived," as one would say. We have to grow cotton as there is no market at all available for getting cash out of other crops beyond the local demand. I am fifty-three years old and want to sell my farm and go somewhere else out West and try my luck. It is not the southern white man's fault. The fault is the system of marketing cotton under distress.[17]

One southern business man caught in the disastrous cotton fall of 1914 felt his inhibitions removed by the debacle. The living conditions of the Georgia cotton farmer have never been more realistically pictured than by Mr. J. T. Holleman, President of the Southern Mort-

---

Quoted by E. V. Wilcox, "The Great White Way of Cotton," *Country Gentleman,* March 31, 1923, p. 18.

[17] F. J. Bivens, *The Farmer's Political Economy,* pp. 13–14. Pamphlet.

gage Company, a Director of the American National Bank of Atlanta, and a "son of the old South." His communication was first published as an open letter to the *Atlanta Constitution* of September 27, 1914, and later republished as a pamphlet:

Pathetic indeed has been the life of the small landowner and tenant farmer in Georgia and the South for fifty years. Courageous, honest, patient, and long-suffering, when shall they see light? When shall their burdens be lifted? In the springtime they go forth with our brothers in black, set their hands to the plow. They bend their backs to the burdens and when the frost falls they have added $1,000,000,000 to the wealth of the world. But small indeed is their share, and meagre is the recompense to them. Every two years, according to the government census, they move from one place to another. They build no homes, they live in rude huts, no flowers about their dwellings, no trees to shade them from the sun, consumed by the summer's heat and chilled by the winter's cold, no lawns about their houses, no garden fences, and with the accursed cotton plant crowding the very threshold of their rude dwellings. . . . Their sons and daughters come to manhood and womanhood, depart from the farm and are lost to them in some distant community. Finally, when their fight is over they are laid to rest in the rude churchyards of the country, others take their place and continue the fight. They have established no permanent home, their kith and kin are scattered far and wide, and the places that knew them once, know them no more, forever. I have no word of criticism for men like these. I know them, I have lived among them. I sprang from them. Who shall undertake to lead these men out of the wilderness of their troubles? Men whom they elevate to high offices in the state and national government are ever ready to teach them politics but they are not prepared to help them solve their problems.

The system of all cotton and no foodstuffs entails more suffering upon the Negroes than upon anyone else. Theirs, indeed, is a struggle for existence. It goes without saying that if anybody shall go hungry and naked, it will be the Negroes.[18]

The Western Cotton Belt has not gone unrepresented in opinions on the living conditions in the cotton industry. In an indignant open letter to the *Dallas News*, T. N. Jones, an attorney of Tyler, Texas, wrote:

> . . . in the South, there is more abject poverty and illiteracy than in any other country on earth in which a high state of civilization is supposed to exist.
>
> The squalid condition of the cotton raisers of the South is a disgrace to the southern people. They stay in shacks, thousands of which are unfit to house animals, much less human beings. Their children are born under such conditions of medical treatment, food, and clothing as would make an Eskimo rejoice that he did not live in a cotton growing country. Without exception around these shacks there are no decent sanitary accommodations. There are no places for the production and care of live stock or poultry. In hundreds of instances there are no arrangements for garden or places where vegetables can be raised.
>
> There is not one landowner in forty who raises cotton and cultivates his own farm who now has his own money in the bank with which to finance entirely the production of his crops for 1928.
>
> There is not one tenant farmer in one hundred throughout the whole South who has his own money on hand with which to finance the production of his crop for 1928.[19]

One must heed the warning that opinions as to how

[18] *Is the South in the Grip of a Cotton Oligarchy?* pp. 17, 18.

[19] *Dallas News*, Feb. 17, 1928.

other people live are as likely to be biased and impressionistic as opinions as to how they should live. Observers of living conditions among any group, unless checked by scientific measurements, are likely to have behind their reports an implied conception of what is right and fitting in the matter of living standards for certain classes and races. We have seen evidences of this in the review of opinions just passed. But the foregoing opinions on how the cotton farmer lives cannot be dismissed as without value. In spite of conflicting sources and viewpoints, the observations possess a remarkable unanimity. The joint report of the Industrial Conference Board and the United States Chamber of Commerce on *The Condition of Agriculture in the United States*,[20] published in 1927, furnishes a brief summary of the opinions stated: "The section which depends on cotton presents the most unsatisfactory aspects. The income of the cotton farmer is on the average very small and the status of these farmers makes most of the South a dark spot in the agricultural picture."

A sampling of opinions from varying sources, such as has been attempted in the foregoing, reaches the conclusion that cotton is a cheap crop, produced by cheap labor. And poor folks have poor ways. "No one crop," said Henry Grady, "will make a people prosperous. . . . Whenever the greed for a money crop unbalances the wisdom of husbandry, that money crop is a curse." "Cotton," writes a country editor in picturesque phrases, "is something to exchange for sowbelly and molasses, some flour and a little coffee, for cotton hose, cotton dresses, and cotton shirts. You can't build consolidated

[20] P. 72.

schools, cement roads, and painted homes with cotton; it isn't the thing to trade for silken hose, automobiles, radios, washing machines, farm lighting systems, and bath tubs. We've got to replace cotton with something more profitable or get off the earth."[21]

STUDIES OF THE COTTON FARMER'S STANDARD OF LIVING

Set over against opinions as to how the cotton farmer lives, we have the studies done by the United States Department of Agriculture and other research groups on rural standards of living. These studies have been undertaken in many of the divisions of the United States, but the returns from southern states are tabulated separately and can be compared with those from other sections. Variable factors in the analysis are the levels of tenure, the races, and the extent to which southern farmers are cotton farmers. Enough separate tabulations have been made for landowners, share renters, and for white and black farmers to show the general trends. The extent of diversification may be judged from the ratio of food furnished by the farm to living purchased. Both of these factors are given and furnish valuable material for study. The variations in living standards between the different areas can be shown by citing studies from the eastern, western, and Gulf states.

In 1926 E. L. Kirkpatrick published the results of a study of the farmer's standard of living carried on over a period of years by the Department of Agriculture in coöperation with twelve colleges and universities in eleven

[21] *Elizabeth City* (N. C.) *Independent,* Oct. 1, 1926. The writer is aware of the delightful impropriety of calling W. O. Saunders a country editor.

states.[22] Only white farmers were studied, and localities were chosen in which average farming conditions prevailed. The data were gathered during the years 1922 to 1924. As will be seen, the study represents a high level of southern farming conditions. Three hundred and seventeen families were studied in New England, 1,439 in the North Central States, and 1,130 in localities representing the dominant type of farming in three southern states: South Carolina, Alabama, and Kentucky.[23] Cash tenants, share tenants, and croppers were included, forming almost 30 per cent of all farmers studied.

The average size of farms reported was 147.5 acres, their average value, $13,788, and the average value of belongings was $1,597.50 per family of 4.4 persons.[24] The New England families owned personal goods to the value of $1,692.20, and southern families ranked last with $1,551. For all families the farm furnished $683.70 worth or 42.8 per cent of total value of all goods; the New England farms furnishing $656.40 or 38.8 per cent of goods consumed by their families as compared to southern farms which furnished $707 worth or 45.6 per cent.[25] Food proved, of course, to be the most important item of the living, making up 41.2 per cent of the value of all goods used. These findings agree with previous studies. The houses occupied by New England families were largest with 9.6 rooms, in comparison with 5.9 rooms per southern household. This gives two rooms per person for New England families and 1.2 per person in the South.[26]

---

[22] *The Farmer's Standard of Living*, Dept. of Agriculture Bulletin 1466.

[23] *Ibid.*, pp. 4–6.

[24] *Ibid.*, pp. 7, 13.

[25] *Ibid.*, p. 15.

[26] *Ibid.*, pp. 21–22.

In New England 52.7 per cent of the families studied, owned and used automobiles, 54.1 per cent of the southern families, and 78.2 per cent of the north central families. Southern farm families had the highest average expenditure for household help and laundry sent out.[27] This item is owing to the presence of cheap domestic help furnished by Negro women. This fact also throws light on the rather high social and economic status of the southern families studied.

Tables of expenditures for personal and advancement goods show interesting rankings. Southern farm families spend more for formal education, church, Sunday school and missions, and tobacco, and less for reading matter, dues of organizations, recreation, gifts, toilet articles, candy, and sodas than any other farm group in the United States.[28] They also spend less for clothing. These studies are highly valuable as offering grounds for comparison between the higher levels of farm families in the various sections.

The value of goods consumed by the Alabama and South Carolina families is divided into living furnished and purchased in the table [29] on the following page.

The South Carolina families spent for their living an average of $1,481.80. Of this amount 47.1 per cent was furnished by the farm. The greatest single item was food, which made up 45.4 per cent of the total budget. It is shown that 74.8 per cent of the food was furnished by the farm. Alabama families had a higher budget of $1,614.80 but with approximately the same proportion furnished by the farm.

The same trends are shown in the averages of the

[27] *Ibid.*, p. 24. [28] *Ibid.*, tables 5 and 6, pp. 25, 26.

[29] Compiled from Bulletin 1466, pp. 16–17.

TABLE X

CONSUMPTION OF GOODS BY ALABAMA AND SOUTH CAROLINA FARM FAMILIES

1922–1924	302 FAMILIES SOUTH CAROLINA AVERAGE SIZE 5.2 (5.6)†			558 FAMILIES ALABAMA AVERAGE SIZE 4.9 (5.3)†		
	Furnished	Purchased	Total	Furnished	Purchased	Total
	74.8%	25.2%	45.4%	76.1%	23.9%	47.2%
Food	$503.70	$169.40	$673.10	$580.40	$182.00	$762.40
Clothing	—	233.40	233.40	—	257.20	257.20
Rent	138.30	—	138.30	125.30	—	125.30
Furnishings	—	25.00	25.00	—	43.50	43.50
Operating (Fuel etc.)	56.50	129.60	186.10	42.10	135.80	177.90
Health	—	46.00	46.00	—	47.00	47.00
Advancement	—	98.20	98.20	—	117.10	117.10
Personal Goods	—	33.20	33.20	—	42.20	42.20
Insurance	—	43.80	43.80	—	40.70	40.70
Unclassified	—	4.70	4.70	—	.90	.90
	$698.50	$783.30	$1,481.80	$747.80	$866.40	$1,614.20
Per Cent	47.1%	52.9%	—	46.3%	53.7%	—

†Average size of household, the family plus relatives and boarders.

budgets of 861 white families in selected localities of Kentucky, Tennessee, and Texas. The total family living cost was $1,436.00 for 4.6 persons, and $536.80 or 36.9 per cent was produced on the farm.[30]

TABLE XI

BUDGETS OF WHITE FAMILIES ON FARMS IN KENTUCKY, TENNESSEE, AND TEXAS

	FURNISHED	PURCHASED	TOTAL
Food	$383.8	$248.0	$631.8
Clothing	——	254.7	254.7
Rent	137.9	2.0	139.9
Furnishings	——	28.5	28.5
Operating	14.9	158.0	172.9
Health	——	67.0	67.0
Advancement	——	84.3	84.3
Personal	.2	16.1	16.3
Insurance	——	36.9	36.9
Unclassified	——	3.1	3.1
	$536.8	$898.6	$1,435.4

When these families are divided into the tenure groups of owners, tenants, and croppers, the following are the average budgets:[31]

TABLE XII

FAMILY BUDGETS BY TENURE

TENURE	PERSONS PER FAMILY	LIVING FUR-NISHED BY THE FARM	PCT. LIVING FUR-NISHED BY FARM	LIVING PUR-CHASED	TOTAL LIVING	PCT. OF OWNERS' LIVING
Owners	4.4	$631.30	38.6	$1003.70	$1635.00	100.0
Share-Tenants	4.8	503.70	36.5	874.10	1377.80	84.2
Croppers	4.9	318.40	33.6	628.50	946.90	57.9

[30] Preliminary Report Dept. of Agriculture, Press Release, Jan., 1925.

[31] *Loc. cit.*

The value of the family living diminishes from owner to cropper as one would expect, although they have the larger families. It is worthy of note that the percentage of goods furnished by farm is lower for the tenants, and lowest for the croppers.

These studies may be said to represent the higher levels of southern farmers. That they practice diversification is shown by the large percentage of supplies furnished by the farm. The living purchased comes from the cash crop, and from the expenditures it is clear that the farms are larger than the average southern holdings. The exclusion of Negroes leaves out of this group a body of cotton producers who would have brought the averages down.

A later study by the Department gives an analysis of cotton farmers on smaller acreage. Gwinnett County, Georgia, is a typical Piedmont cotton-producing area, inhabited mostly by white farmers. The average cropper cultivates twenty-four acres; the average tenant twenty-six, and the average landlord, with his tenants, has forty-eight acres in cultivation. From the 288 farms studied, it was found that the typical farm had eight acres in cotton, thirteen in corn, and four in other crops.

The standard of living study of these 288 white farm families showed an average net cash income in 1924 of $424 per family of five. Their cash income was supplemented by food, fuel, and shelter from the farm to the value of $396 per family, giving a total of $820 a year. Of their net cash income they had to pay ordinary living expenses averaging $291 per family, leaving $133 to be applied to their debts or added to their capital. Most of these farmers were tenants. Only one work animal was found on 40 per cent of the farms and only 14 per cent had over two. These people have meager wants, but their wants exceed or equal their income. The average annual expenditure of each family for books, magazines, recreation, amusements, education, and religion was $24 per year. It was almost equaled by the purchase of their only luxuries, $15 per year for tobacco and snuff. Thirty-four per cent of the families live in houses of only one thickness of lumber. Their wants would be even more scantily supplied but for the fact that 90 per cent of the housewives have cheap sewing machines and make part of the clothing for the family. The exodus of young people from these farms is increasing. Forty-seven per cent of owners' children who have grown up had left the farm to go into some other occupation besides agriculture. However, only 29 per cent of the croppers' children leaving home have deserted agriculture. The Department of Agriculture estimates that there are perhaps 100,000 families along the border of the old Cotton Belt who live no better than those of Gwinnett County.[32]

Even lower levels of living were found for Negro farmers in Kentucky, Tennessee, and Texas.[33] Six hundred eleven dollars and ten cents was found to represent the living of an average family of 4.8 people. Of this amount $239.70 was furnished by the farm and $371.40 purchased. The following table shows how the expenditures were distributed:

---

[32] Dept. of Agriculture Press Release, April 11, 1927.

[33] Dept. of Agriculture Preliminary Report, Press Release, Jan., 1925.

TABLE XIII

BUDGETS OF NEGRO FAMILIES IN KENTUCKY, TENNESSEE, AND TEXAS

	FURNISHED	PURCHASED	TOTAL
Food	$178.6	$148.1	$326.7
Clothing	—	107.1	107.1
Rent	41.0	—	41.0
Furnishings	—	4.5	4.5
Operating	20.1	35.7	55.8
Health	—	24.8	24.8
Advancement	—	28.3	28.3
Personal	—	8.9	8.9
Insurance	—	14.0	14.0
	$239.7	$371.4	$611.1

The Negro families also show variations according to tenure levels. The following tabulation is arranged to show this:

TABLE XIII-A

TENURE	PERSONS PER FAMILY	FURNISHED BY FARM	PCT.	PURCHASED	TOTAL LIVING	PCT. OF OWNERS' LIVING
Owners	4.8	$304.50	44	$378.10	$682.60	100
Tenants	4.9	273.70	40	398.80	672.50	98
Croppers	4.7	186.00	34	350.30	536.30	78

The decrease in goods furnished by the farm shown by croppers may be regarded as the diagnostic sign of the one-crop system with cotton for cash. Unlike the white farmers, black landowners and share tenants as a class show little difference in income.

It is desirable to present one study of the older eastern and another of the newer western cotton area. One of the most thorough and painstaking investigations made of living conditions of the eastern cotton farmer was carried out by the North Carolina Tenancy Commission in 1922. The data are presented by C. C. Taylor and C. C. Zimmerman in *Economic and Social Conditions of North Carolina Farmers*. The survey included 1,014 farm families divided between three counties typical of the Mountain, Piedmont, and Coastal Plain sections. In the Coastal Plain county, the 339 farms studied had 93.6 per cent of their crop land in cotton, corn, and tobacco; in the Piedmont the 335 farms surveyed had 81.4 per cent in these three crops. As the mountain farms do not grow cotton they are omitted. The data were analyzed both in regard to race and status of tenure: operator-landlords, owner-operators, tenants, and croppers. Incomes ranged all the way from $2,385.85 for white landlords to $640.59 for Negro croppers in the Coastal Plains. In the Piedmont the range for the same groups is from $994.70 to $208.60. The highest group received $1.25 per day per member of the family; the lowest barely ten cents.[34]

Table XIV shows the average cash income these investigations show per family of five for 1921.

In the Coastal Plains 23 per cent of the 339 farmers studied were landowners; in the Piedmont 54 per cent of 335 families owned their farms. In the Coastal Plains 59 per cent of the families studied were white; in the Piedmont, 56 per cent. Renters averaged 5.42 members per family in the Piedmont and 5.05 in the Coastal Plains, while the landowner had 4.17 and 5.37 persons per family. The average number of acres cultivated per family was slightly less than eighteen; for renters it was

[34] Taylor and Zimmerman, *Economic and Social Condition of North Carolina Farmers*, p. 27. Report of State Tenancy Commission, 1922.

TABLE XIV

CASH INCOME OF NORTH CAROLINA FARM FAMILIES

	Landlord Operators		Owner Operators	
	White	Black	White	Black
Coastal Plains......	$2,385.85	$1,996.00	$1,129.50	$1,670.00
Piedmont...........	994.70	379.00	604.17	536.15
	Tenants		**Croppers**	
	White	Black	White	Black
Coastal Plains......	$854.15	$711.07	$700.84	$640.59
Piedmont...........	336.35	374.24	364.74	208.60

slightly over sixteen. The higher cash incomes for the Coastal Plain farmers are due to the larger acreage in cash crops, cotton and tobacco.

This gross income, however, is supplemented by food raised on the farm. This ranged from around $500 for owners to around $200 for croppers. (See Table XV.)

None of these families, averaging over five members, consumed as much as two dollars worth of food per day, less than forty cents per person. There were six groups among tenants and croppers that consumed food to the value of less than a dollar per family per day. It is again noted that "the per cent of family food supply raised on the farm progressively declines from landlord to owner to tenant to cropper."[85]

Some items in the diet are worthy of consideration. For instance, the production of milk per day per person reached as high as a quart only for the white landlord group. For the white farmer in general, it ranged below

[85] *Ibid.*, pp. 17, 18.

TABLE XV

LIVING RAISED ON FARM AND BOUGHT

COASTAL PLAIN	OPERATOR–LANDLORD		OWNER–OPERATORS		TENANTS		CROPPERS	
	White	Black	White	Black	White	Black	White	Black
Food Raised	$400	$514	$386	$474	$323	$200	$194	$156
Food Bought	56	82	131	134	93	113	96	110
Food Total	$456	$596	$517	$608	$416	$313	$290	$266
Per Cent Raised	87.7	86.2	74.7	78.0	77.6	63.8	66.9	58.6
PIEDMONT								
Food Raised	$517	$323	$467	$346	$328	$198	$257	$172
Food Bought	71	82	154	186	62	115	98	114
Food Total	$588	$405	$621	$532	$390	$313	$355	$286
Per Cent Raised	87.9	79.7	75.2	65.0	84.2	63.2	72.4	60.2

.4 of a quart; for Negro renters[36] from .07 to .6. The production of pork ranged from 600 to 200 pounds per family.[37] The data on housing is also significant. The average landless farmer had 3.74 rooms in his house as compared with 4.61 for owners. The average white farmer had 4.38 rooms in his house as compared to 3.72 for the Negro farmer. Landowners had 1.07 persons per room as compared with 1.38 persons per room for landless farmers. White homes had 1.11 persons per bed as compared to 1.52 for Negro homes. Lastly, 17.6 per cent of landless farmers lived in one or two-room houses, and 14.4 of all whites and 13.9 per cent of all Negroes lived in homes of two rooms or less.[38] Not a Negro family nor any white tenant or cropper family in the area surveyed had running water, an indoor toilet, or a bathtub in his home. In the landless families over 31 per cent of the fathers and mothers could not read. The average cropper had attained only to the third grade in school, while the average Negro farmer of all ranks had received less than a first grade education. This exhaustive survey of the owners, tenants, and croppers in cotton and tobacco-growing areas shows with the realism of statistics the low living standards in the old Cotton Belt. It is true that 1921 was a bad year for cotton, but all of the conditions presented in the survey must be regarded as the cumulative result of years of marginal incomes.

A study made by J. T. Sanders[39] for 1919 and published by the Department of Agriculture in 1922 devoted itself to the analysis of living conditions of 368 cotton farmers in six counties of the Black Prairie of

---

[36] *Ibid.*, p. 19. [37] *Ibid.*, p. 20. [38] *Ibid.*, p. 41.

[39] *Farm Ownership and Tenancy in the Black Prairie of Texas*, Dept. of Agriculture Bulletin 1068.

Texas. In this region the change from a grazing economy to cotton farming has been unusually rapid and the development of the tenant system has kept pace. The historical background of the plantation system was absent in the Black Lands, and cotton farming has accordingly been based on an infiltration of white tenants. According to the 1920 Census the percentage of Negro farmers was still low—15.9 per cent.[40]

The standards of living as shown by the total cost of family living for 368 unselected farm operators was $965 for share croppers, $1,243 for share tenants, and $1,742 for owners. The average size of families was found to be about the same, so that the living cost of the cropper family can be estimated at 55 per cent, and the average share tenant's living at 71 per cent of the cost of family living of owners.[41] Of all living expenses croppers have $262 furnished by the farm, share tenants $424, and owners $575. Thus croppers receive only 41 per cent, and share tenants 75 per cent as much family living from the farm as owners. Croppers, thus with the lowest standard of living, $965, buy the most groceries, $310, to $296 for tenants and $294 for owners. The figures are not adequate in that they do not show the wide variations in knowledge of the selection, preparation, use, and qualities of the articles of diet. Many of these articles are bought at country stores, where often no provision is made for an adequate supply of milk, vegetables, and fruit.

The average amount spent for recreation, education, and advancement goods is strikingly small for all classes. Ten to fifteen dollars per family per year is spent for

---

[40] *Ibid.*, p. 1.

[41] *Ibid.*, pp. 53–54.

recreation. "Few families of any tenure class take vacations and but few more patronize movies or theaters." It is interesting to note that approximately twice as much is spent for tobacco and similar personal expenses as for recreation.[42] About one out of six croppers own automobiles to one out of two share tenants and three out of four owners. Telephones cost less and are relatively more in use. About one out of five croppers, half the share tenants, and six to seven out of ten owners have telephones.[43] All owners read periodicals and daily papers, but 39.1 per cent of croppers reported no periodicals whatever.[44]

In the schools it was found that 96.5 per cent of owners' daughters were promoted for the school year as compared with only 77.2 per cent of tenants' daughters; and 88.6 per cent of owners' sons were promoted as compared with only 65.6 per cent of tenants' sons. As Sanders states, "the tenant's child is from six months to a year behind the owner's child in grade attainments."[45] The enrollment of tenants' children reaches its lowest stage during cotton-picking time in October, November, and December. "This low enrollment . . . is due to the fact that tenants as a rule feel that they cannot afford to hire their cotton picked and to the fact that the landlords expect, and sometimes demand, that renters' children be put into the cotton fields in order to rush picking as much as possible." The low enrollment for cotton tenants' children during other parts of the year is due to "the fact that children of tenants who move into the school district are not enrolled the first part of the school

---

[42] *Ibid.*, p. 54.

[43] *Ibid.*, p. 55.

[44] *Ibid.*, p. 56.

[45] *Ibid.*, p. 59.

term, while those moving out are not enrolled during the latter part of the term."[46]

The relation of living purchased to the cash receipts from the cotton farm affords a problem full of interest and difficulty. It has been shown that the living furnished the farm family ranges around 40 per cent of the total expenditures and is often less for the lower levels of tenure. For each family, the food, fuel, and shelter from the farm home tend to be a constant figure, while living purchased varies with the fluctuations of agricultural prosperity. Thus in 1918 and 1919, years of good farm prices, the family living from the farm was only one-fifth of the total family income for 2,967 families studied; in 1921 and 1922, years of agricultural depression, the family living was about one-third of the total family income from 3,597 farm families studied by the Department.[47] Living from the farm thus serves as a balancing factor to conserve in some measure the family's standards during depressions. The wide fluctuations in the price of cotton, it can be seen, give the extent of diversification, on which the living from the farm depends, a most important bearing on the standard of living.

Mr. E. A. Goldenweiser[48] in 1916 estimated from a statistical analysis of the 1910 Census returns that the labor income of the average American farmer was about $600, $200 of which he received in cash, and $400 he secured in goods from the farm.[49] The estimate was based on a gross farm income of $1,236 minus $512 paid for

[46] *Ibid.*, p. 58.

[47] H. W. Hawthorne, *The Family Living from the Farm*, Dept. of Agriculture Bulletin 1338, p. 12.

[48] *The Farmer's Income*, Farmers' Bulletin 746, 1916.

[49] *Ibid.*, p. 1.

farm expenses such as fertilizers, feed for stock, labor, etc., and $322 interest at five per cent on the farm capital.[50] This was above the average of urban wage-earners and below that of salaried employees.[51] An analysis of 4,018 farms selected from regions representing better than average conditions showed the average earnings of these farm families to be $952 a year, of which $400 came from the farm as food, fuel, and housing.[52] In this case also, the farm living remained the constant factor.

Allowing for the rise in food prices these figures agree with the figures on the value of living furnished from the farm, cited for the higher groups of farmers. In these cases, also, the living from the farm tended to remain the constant factor. Diversification may be said to approach its limit when it feeds the family and the work animals. Any further rise in the standard of living will depend on increased production of cotton, or other cash crops.

Thus on even the better diversified farms the food crops may do no more than feed the household and work animals, leaving the living purchased to be defrayed by the receipts from cotton. The retarded development of urban and industrial centers in the South has not created wide markets for truck and dairy products. Accordingly, in many instances the farmer's money income will depend on two factors: the acreage in cotton and the level of tenure. Acreage in cotton determines how much the farmer produces, and his tenure level determines what part of his product he receives. As a matter of fact this formula for ascertaining the money receipts of the cotton farm is an oversimplification in that it

---

50 *Ibid.*, pp. 3–4.

51 *Ibid.*, Fig. I, p. 6.

leaves out the ever-present fluctuations in yield and price.

Morris Sheppard,[53] former United States Senator from Texas, suggested in 1924 an estimate of the average money income to the cotton grower for the sixty year period since the Civil War. For that period he figured the average yield of lint per cotton acre at 178 pounds, the average farm price at 13 cents per pound, and the average income per acre at approximately $23.05. Setting the average cotton patch at seventeen acres, Sheppard estimated the average gross income at slightly less than $400 for the whole farm family. The 1920 Census of Agriculture showed the average cotton acreage per farm reporting cotton to be 17.7 for the United States; slightly over fourteen acres for the Eastern Belt, about twelve acres for the Gulf states, almost twenty-five acres for the western states, and thirty-three acres for Texas.

In this general way the amount of cash income may be determined from the cotton acreage. Then the share going to the producer may be estimated on the basis of tenure: one-half to the cropper, two-thirds to three-fourths to the share tenant, and all, less credit charges, to the owner. We shall have more to say about the credit charges later.

Dr. E. C. Branson and J. A. Dickey [54] made a survey of the money income of 329 farm operators in Chatham County, North Carolina, for 1921. In this region of mid-state Carolina, cotton and tobacco are the money crops, while corn, wheat, oats, potatoes, milk and butter, poultry and pork are the homegrown staples of existence.

---

[53] "Who Gets the Cotton Money?" *Country Gentleman,* Aug. 23, 1924.

[54] *Home and Farm Ownership,* North Carolina Club Yearbook, 1921–22, pp. 59–99.

They found the average cash income to be $424 per family, divided by races and tenure as follows:[55]

TABLE XVI

MONEY INCOME OF NORTH CAROLINA FARM FAMILIES

	Family Cash Income Per Year	Per Cent of White Owners' Income	Daily Cash Income Per Person
135 white owners	$626	100.0	34 cents
41 black owners	597	95.6	32 "
66 black renters	289	44.4	16 "
38 white renters	251	40.1	14 "
36 black croppers	197	31.4	10 "
13 white croppers	153	24.2	8 "

The share tenants were found to have a cash income less than half that of owners, and croppers received less than a third. The Negroes tended to outrank the white tenants and croppers. The variable factor we most need to know in this connection is the size of the farms.

It is necessary to emphasize in regard to the cotton farmer two facts: first, the extreme concentration of the farm's cash income on cotton, and second, the fluctuation in living purchased that results from the variations in price of cotton. It is difficult to overestimate cotton's place in southern agriculture as a cash crop. For example, studies[56] have shown that farms in Catawba County, North Carolina, with only 6 per cent of their tillable land planted to cotton, draw 35 per cent of their total cash receipts from the sale of lint and seed; in Jones County, Mississippi, 8 per cent of the land in cotton accounts for 38 per cent of the total income; while

[55] *Ibid.*, p. 66. [56] Hawthorne, *op. cit.*, Table I, p. 4.

in Sumter County, Georgia, 23 per cent of the farm in cotton brings in 79 per cent of the cash receipts. The farmer who grows cotton at all is thus dependent upon it to a most remarkable extent for the amount of the family living that may be purchased.

The preceding chapters on hazards of production and of the cotton market show the great fluctuations in cotton yields and incomes. Table XVII showing the average per acre value of cotton to the producer has been translated into terms of annual cash incomes for family-sized cotton farms of ten and twenty acres. The purchasing power of cotton per pound and per acre in terms of the Bureau of Labor index is also given.

It was the conclusion of the Joint Commission of Agricultural Inquiry of the U. S. Congress in 1921 that "as expressed in purchasing power cotton is relatively less profitable than either corn or wheat."[57]

The reactions of these fluctuations on the farmer's budget deserve more than a passing word of comment. Students of the budgets of workingmen have pointed out the bad effect of alternating periods of prosperity and depression on the standards of living of workingmen's families. No regular planning for expenditures, it is held, is possible; extravagance dissipates the surplus during good times, and hard times can be met only by reducing standards below minimum levels of health and decency. It is likely that an income striking an average between the high and low levels would lead to progressively rising levels of consumption in clothing and diet.

The relations that the hazards of the cotton industry hold to the standard of living of the producer are fairly

---

[57] *The Agricultural Crisis and Its Causes,* Part I, p. 165.

TABLE XVII

AVERAGE INCOME OF 10- AND 20-ACRE COTTON FARMS WITH RELATIVE PURCHASING POWER OF COTTON PER ACRE AND PER POUND—1880–1927

	Average Income for 10 Acres in Cotton	Average Income for 20 Acres in Cotton	Relative Purchasing Power of Cotton Per Pound (1910–1914 = 100 *)	Relative Purchasing Power of Cotton Per Acre
1927	$312.10	$624.20	—	—
1926	208.70	417.40	—	—
1925	317.90	635.80	—	—
1924	372.60	745.20	—	—
1923	423.40	846.80	186	127
1922	351.40	702.80	138	104
1921	211.10	422.20	105	69
1920	202.20	404.40	70	67
1919	606.20	1,212.40	145	125
1918	460.00	924.00	124	105
1917	462.80	925.60	138	117
1916	320.80	641.60	119	99
1915	201.00	402.00	95	85
1914	149.10	298.20	64	71
1913	232.60	465.20	112	108
1912	238.30	476.60	108	110
1911	190.80	381.60	85	95
1910	253.20	506.40	128	103
1909	225.50	451.00	125	103
1908	192.70	385.40	87	89
1907	202.10	404.20	102	97
1906	201.70	403.40	94	100
1905	230.40	460.80	110	109
1904	205.10	410.20	94	103
1903	209.00	418.00	111	103
1902	167.60	335.20	80	78
1901	154.80	309.60	79	72
1900	180.10	360.20	104	106
1899	149.90	299.80	81	79
1898	107.80	214.60	72	78
1897	124.20	248.40	89	81
1896	127.10	254.20	91	85
1895	139.30	278.60	100	80

* G. F. Warren and F. A. Pearson, *The Agricultural Situation*, Table XVIII, p. 122. More recent figures are not yet available due to change made in Bureau of Labor index. Letter from W. F. Callander, Crop Reporting Board, April 5, 1928.

TABLE XVII (*Continued*)

	Average Income for 10 Acres in Cotton	Average Income for 20 Acres in Cotton	Relative Purchasing Power of Cotton Per Pound (1910–1914 = 100)	Relative Purchasing Power of Cotton Per Acre
1894	$119.60	$238.20	61	59
1893	137.20	274.40	87	65
1892	147.80	294.60	99	105
1891	144.80	289.60	84	76
1890	152.20	304.40	96	90
1889	191.50	382.00	94	76
1888	184.20	368.40	94	86
1887	177.50	355.00	94	87
1886	167.40	334.80	90	77
1885	154.00	308.00	94	78
1884	163.00	326.00	99	77
1883	170.80	341.60	91	76
1882	231.70	463.40	88	83
1880	235.80	471.60	96	89

well known. E. E. Miller reports a country banker as observing that "the farm deposits begin coming into his bank in the fall of a good year, piling up until about January, then decreasing until spring, when the borrowing season begins, and finally giving away by midsummer almost altogether to farmers' notes."[58] Replies to questionnaires[59] sent out indicated that of the money received for the 1925 crop, 11.6 per cent would be spent for improvement, new machinery, buildings, and better live stock, 5.4 per cent would be placed in sound investments, and 17 per cent would be wasted.

The unwarranted expenditure of the higher cash incomes for cotton is found to a greater degree among many of the poorest producers. An investigator of the

[58] "Cotton a National Crop," *Review of Reviews*, July, 1926, p. 72.
[59] *Loc. cit.*

living conditions of the cotton farmers of North Carolina in 1916 and 1917, years in which they had received high prices, says that in times of high cotton prices, "the farmer soon falls victim to the hordes of agents who radiate from the small towns and have just the thing the farmer needs, usually at exorbitant prices." [60] Automobiles, nostrums, horse doctor books, enlarged family portraits, expensive family Bibles, and large wall maps of the state and the nation are among the many things the farmer buys. "Because of lack of thrift and his ignorance, he soon parts with his money, and the essentials for the farm and the family are still lacking." [61] The student of the subject will agree with E. E. Miller when he says: "No part of the country needs the discipline of systematic saving or a realization of the importance of a little accumulation as does the rural South," [62] and will be tempted to add that no section has less opportunity to learn these principles as long as its cash income, dependent on cotton, is subject to violent fluctuations.

Yet another phase of family living purchased by the cotton farmer remains to be discussed—the vexatious problem of credit. It is difficult to state how many growers run accounts at supply stores, commissaries, or with landlords, but estimates previously cited run from 50 to 90 per cent.

Studies made in the field, cited in the chapter on the cotton system, seem to point to an interest charge on short-time merchant credit of 20 to 25 per cent per annum. If this figure is accepted as reasonable it is

---

[60] W. H. Swift, *Child Welfare in North Carolina*, National Child Labor Commission, 1918, p. 170. [61] *Loc. cit.*

[62] *Op. cit.*, p. 72.

worth while to examine the bearing such a charge has on the farmer's standard of living. A likely procedure in arriving at real expenditures would be as follows: In the budget of farmer's living during the growing season on consumption credit, the items headed "family living purchased" should be credited to the family living at from 75 to 85 per cent of their money value. In other words, if the family while making the crop ran an account of $100 at a supply store, they may be considered as having consumed $75 to $85 worth of goods. The other has been paid for interest charges. To this extent the figures given us on livings of southern farmers are padded with high interest charges that do not represent goods consumed. For example, the average amount of credit advanced to 1,330 Negro families on cotton plantations in 1920 and 1921 was previously cited at $289 for croppers and $555 for tenants. It may be safely assumed that these families consumed goods to the value of from 75 to 80 per cent of the sums mentioned, the remaining being interest charges. This fact must be regarded as of particular importance, because the lower income groups, such as the croppers, tend to buy a large percentage of their food and are less able to pay cash for it. It is thus extremely likely that their living standards are even lower than the low figures secured suggest. It is desirable that future standards of living, studies among farmers in the South take this factor into account and estimate the percentage of the living that is bought on a time basis. At that, such a study will leave out of consideration the poverty of resources of the small town store in supplying a varied and healthful dietary. It may also be suggested that clothing cost will be found less likely in the Cotton Belt to come under merchant credit. The purchasing of needed articles of apparel is deferred until the cotton is sold. It is not uncommon in the fall to see a tenant's whole family invade the stores for shoes all around. The period after harvest at which credit relations are resumed by tenants depends upon their surplus after making settlement. If the yield has been poor and the price low, it will be necessary for them to resume credit relations at once. In such cases practically the whole living purchased will be subject to time prices.

THE FAMILY LIVING FROM THE FARM

An evaluation of the family living from the cotton farm is faced with an equal number of perplexities. Its important place cannot be denied. Food, fuel, and shelter come high to urban dwellers, and it simply adds to the farmer's standard of living if he can secure them partly as by-products of his occupation. In estimating these services of the farm home the Department analysis placed the value of food and fuel furnished halfway between farm valuation and city prices. The rent was estimated at 10 per cent of the current valuation of the house.

We have here again to deal with the extent of diversification. It has already been shown that tenants and croppers not only have a smaller amount of living furnished by the farm but a smaller per cent than owners. It has also been shown that, in general, owners diversify most and croppers least. To the landlord the tenant's farming operations are a financial not an agricultural venture. He can be expected to exact cash crops. In the Texas survey J. T. Sanders[63] found that owners had

[63] Dept. of Agriculture Bulletin 1068, p. 19.

about six-tenths of their land in cotton, share tenants over two-thirds, and croppers about four-fifths. In the North Carolina study [64] it was found in the Coastal Plains that owners had 2.6 per cent, landlords 7.5 per cent, tenants 10.5 per cent, and croppers 20.3 per cent of all their tillable land in cotton.

Accordingly, as heretofore in the study of living standards, we shall be unable to strike averages that are general. Instead we shall present the high and low levels of living. An analysis [65] of the family living from 1,008

TABLE XVIII

FAMILY LIVING FROM FARM, NORTH CAROLINA, GEORGIA, MISSISSIPPI

	No. Farms	Value of Food	Cords Wood	Value
North Carolina, Catawba Co., 1918	304	$453	11	$26
Georgia, Sumter Co., White "	336	469	16	42
Negro "	214	374	13	38
Mississippi, Jones Co. 1919	154	515	7	27
	1,008			
Averages U. S.	7,738	$320	4	$14

	Value House Rent	Value All Items	No. of Adult Units	Value Per Adult
North Carolina, Catawba Co., 1918	$ 89	$573	5.2	$110
Georgia, Sumter Co., White "	179	690	4.5	153
Negro "	59	471	5.9	80
Mississippi, Jones Co. 1919	85	627	5.2	121
Averages U. S.	$184	$518	4.1	$130

[64] Taylor and Zimmerman, *Economic and Social Conditions of North Carolina Farmers*, p. 12.

[65] Hawthorne, *op. cit.*, pp. 8–9.

farms in Catawba County, North Carolina, Sumter County, Georgia, in 1918, and Jones County, Mississippi, in 1919 gave interesting results. The value of all items furnished from the farm ranged from $471 for 214 Negro families to $690. The values per adult were $80, $110, $121, and $153 per year. The southern farmers in spite of larger families exceed the averages for the county in value of food and fuel but were excelled in value of house rent and total value for each adult.

These are exceptional southern farms in well balanced agricultural areas. The food produced per farm is thus tabulated and compared with averages for 7,738 farms in selected localities in the United States.[66]

Southern farms exceed in the production of corn, potatoes, sirup, butter, milk, pork, and vegetables. They are exceeded by the United States averages in beef, poultry, and eggs. (See Table XIX.)

Table XX (page 246), represents the average annual food consumption of 255 farm families located in North Carolina, Georgia, and Texas. It shows many of the characteristics of the southern dietary in the predominance of pork over beef, but is rather noticeable in its omission of collards and the use of more flour than corn meal.[67]

There exists plenty of evidence to show that on the vast majority of southern farms such a dietary remains an unattained ideal. Typical of much Cotton Belt farming are the conditions found by the *Augusta (Georgia) Survey* to be true of its county, Richmond. With the

---

[66] *Ibid.,* Table 2, p. 8.

[67] C. L. Goodrich, from *Producing Family and Farm Supplies on the Cotton Farm,* Farmers' Bulletin 1015, p. 4.

TABLE XIX

QUANTITIES AND VALUE OF FOOD FURNISHED BY FARMS

	Corn		Wheat		Potatoes		Sirup		Fruits Vegetable	Butter	
	Bu.	$	Bu.	$	Bu.	$	Gal.	$	$	Lbs.	$
North Carolina, Catawba Co., 1918	13	19	35	81	27	31	12	9	52	90	31
Georgia, Sumter Co., White 1918	16	28	16	37	30	30	22	24	44	91	33
Georgia, Sumter Co., Negro 1918	28	49	8	18	29	29	24	25	29	65	20
Mississippi, Jones Co., 1919	14	25	—	—	49	49	42	50	37	156	73
Average U. S.	2	4	4	9	16	16	5	5	48	103	42

	Milk		Beef		Pork		Poultry		Eggs		Misc.
	Gals.	$	Lbs.	$	Lbs.	$	No.	$	Doz.	$	$
North Carolina, Catawba Co., 1918	289	57	24	5	548	123	33	25	60	20	5
Georgia, Sumter Co., White 1918	264	47	5	1	895	172	53	28	80	24	1
Georgia, Sumter Co., Negro 1918	204	35	9	2	734	139	26	13	44	13	2
Mississippi, Jones Co., 1919	341	92	—	—	528	106	21	11	62	27	2
Average U. S.	284	59	56	6	471	71	33	23	118	36	1

TABLE XX

AVERAGE ANNUAL CONSUMPTION OF VARIOUS ARTICLES OF FOOD PER ADULT PERSON BY 255 FARM FAMILIES IN NORTH CAROLINA, GEORGIA, AND TEXAS

ARTICLE	UNIT OF MEASURE	AMOUNT CONSUMED PER ADULT PERSON
Vegetables:		
Beans, green	Pecks	5.7
Beets	do	3.5
Cabbages	Heads	14
Cucumbers	Pecks	2
Melons	Number	15
Onions	Pecks	1.5
Peas	do	1.13
Potatoes (Irish)	Bushels	2
Potatoes (sweet)	do	5.11
Sweet corn	Dozen	6.8
Tomatoes	Pecks	4.1
Turnips	do	4.5
Fruits:		
Apples	Bushels	1.4
Pears	do	.3
Peaches	do	1.5
Grapes	Bushels	.25
Berries	Quarts	11.5
Cereals:		
Corn meal	Pounds	136
Flour	do	224
Sirup	Gallons	4
Sugar (54 pounds)	do	8.2
Dairy products:		
Butter	—	—
Butter milk = milk	Quarts	482
Milk	—	—
Beef	Pounds	12
Pork and lard	do	138
Poultry products:		
Poultry	do	57.5
Eggs	Dozen	28.4

possibility of creating markets for their products in Augusta, Richmond County farms produced on an average food per person on the farm as follows:

TABLE XXI [68]

Milk	1/4	pint per day
Butter	1	ounce per day
Chickens	1/2	per month
Eggs	1/2	per day
Pigs	2/3	per year
Beef	1/10	per year
Sirup	8/10	gallons per year
Irish potatoes	9	pounds per year

In the dietary of families on the lower levels of living, pork comes first in value of all the items of food supplied directly from the farm, often amounting to 40 per cent.[69] Corn is the staple article of diet throughout the whole Cotton Belt, prepared as roasting ears, hominy, grits, and meal for corn bread. Molasses, homemade from sorghum and ribbon cane, furnishes the sugar in the diet during the winter months. During the summer, beans, tomatoes, Irish potatoes, cabbage, okra, field peas, and onions are raised in the garden. During the winter months cotton farmers subsist on turnips, collards, sweet potatoes, and fried pork. An observer says of cotton tenants in Oklahoma: "Very few have vegetable gardens of any description. Their supply of meat, milk and butter they must buy or go without—most of the time the latter—and a great many of them are actually on the border line of starvation. The writer has been in their homes when there was not enough on the table for even a smaller family." [70] At best the diet is composed of fried foods,

---

[68] *Augusta Survey*, p. 169. Pamphlet.

[69] Hawthorne, et. al., *Farm Organization and Farm Management in Sumter County, Georgia,* Dept. of Agriculture Bulletin 1034, p. 37.

[70] C. E. Gibbons, *op. cit.*, p. 38.

heavy in fats and starches with a deficiency in milk, eggs, fruits, and green vegetables.

The researches of Joseph Goldberger, Surgeon-General, United States Public Health Service,[71] and others have shown the relation between an insufficient dietary and pellagra. In tenant areas during hard times the cotton farmer's diet tends to be restricted to the "Three M's," meal, molasses, and white meat. "This basic diet," writes Dr. Goldberger, "when made up in conventional proportions is pellagra producing. It contains some vitamin, P-P, derived from the corn meal, dried beans, and collards, but ordinarily this is much too little to prevent pellagra." [72] This diet, it is believed, will lead to the development of pellagra in 40 or 50 per cent of those partaking of it within three to six or eight months.[73] Unfavorable cotton years have been followed by an increased incidence of pellagra [74] as in the well-known prevalence of the disease in 1915 following the depression of cotton values, and in 1927 in the areas affected by the Mississippi flood. Nesbitt,[75] public health officer of Wilmington, has shown that in New Hanover County, North Carolina, from 1912 to 1915, a period of decline in cotton prices, the death rate per 100,000 from pellagra increased from 21.38 to 64.60. During the same period, however, cases of typhoid and communicable disease showed a decrease. In the four cotton raising states on the Mississippi, the deaths from pellagra increased

[71] *Pellagra: Its Nature and Prevention,* Public Health Report Reprint 1174.

[72] *Ibid.,* pp. 5–6.

[73] *Pellagra in the Mississippi Flood Area,* Public Health Reprint 1187.

[74] *Op. cit.,* pp. 5–7.

[75] Cited in Lovall, et al., *Negro Migration in 1916–1917,* pp. 26–27.

from 1,020 in 1924 to 1,850 in 1926, and Dr. Goldberger[76] estimated from 2,300 to 2,500 during 1927. The cases during the same period, it is estimated, increased from 20,000 in 1924 to 45,000 or 50,000 in 1927. Crop failures due to the overflow thus operated as a causal factor.

Much has been done and is still being done to improve dietary habits in cotton areas, but many of the white and black cotton growers live, during periods of low prices, dangerously near the level of John Charles McNeil's "Tar Heel":

> "Oh, I gits my stren'th frum white-side meat,
> I sops all de sorghum a nigger kin eat,
> I chaws wheat bread on Saddy night,
> En Sunday's when my jug gits light."

Some figures on housing have been given in connection with living furnished by the farm. The houses of the higher level groups in the South are not equal to those of New England and the Middle West in regard to water supply, electric lighting, and home conveniences. Greater differences exist in housing within the Cotton Belt. A survey[77] in a Georgia county found the value of the white tenant's house to be about half that of the white owner's; the houses of black owners were valued at 46 per cent of those of the white owners, and the Negro tenants at 26 per cent. For the same groups the black tenants had the greatest number in the family, and the white owners the smallest.

Descriptions of these homes are to be met with in

---

[76] *Pellagra in the Mississippi Flood Area,* p. 5.

[77] Dept. of Agriculture Bulletin 1034, p. 36.

many different reports. In the old Cotton Belt the landlord, in many cases according to an observer, lives in "a big comfortable farmhouse with a generous brick fireplace at each end—the traditional southern home with its large cool rooms, deep verandahs, fine trees, sturdy old scuppernong vines, and in the distance well-kept cotton fields." [78]

The tenant usually lives in an "unpainted, clap-boarded cottage of four small rooms" or less, "ceiled inside" papered with old newspapers, "often with no shade around the house." The yard is "a hot, sandy, little plot of ground" with a dug well, usually open and unprotected. Often the house is unscreened and open to the flies, gnats, and mosquitoes.

A Negro renter lives in "one, two, or three rooms." The cabin is "hot in the summer" and in the winter "almost impossible to heat." Daylight shows between the cracks; the cabin "leaks in stormy weather, and leaves the floor damp for a day or two afterward."

The demands of the Renters' Union of McClain County, Oklahoma,[79] may be regarded as a kind of minimum standard of health and decency set up by tenants in the Western Cotton Belt. The specifications called upon landlords to furnish tenants houses with not less than two rooms and a lean-to. "The said two rooms shall not be less than 14 feet square with a ceiling not less than 8½ feet high. The said room shall be plastered and have a lumber floor." The building shall have at least six win-

[78] F. S. Bradley and M. A. Williamson, *Rural Children in Selected Counties of North Carolina,* Children's Bureau Publication 33, p. 23.

[79] *The Land Question in the Southwest,* Report of Commission on Industrial Relations, X, 9067.

dows, with sashes that can be raised and lowered, and the doors and windows shall be screened. "There shall also be built to said building a front porch at least 16 by 6 feet, which may be roofed with boards and batten." The demands included a stable for at least three animals, a tool shed, and "a chicken coop not less than 10 by 12 feet and 6 feet high."

On the whole it is a rather dreary picture and somber group of facts that we have presented here. It is possibly true that we cannot speak of the average cotton farmer. Conditions vary among the different areas. We have found western cotton farmers making money, for instance, during years in which eastern farmers lost. The varying size of the farm is another factor for which we need more data. The trends as regards the different levels of tenure are unmistakable. It is undeniable that cotton croppers live below what may be called the rural poverty line. However unprepossessing the conclusions are as to the cotton producer's standard of living they appear in substantial agreement with findings of the Industrial Conference Board:

> The average cotton farmer has hardly more than about 25 acres in cotton and as crop diversification is still very rare among cotton farmers, it follows that the total gross cash income of these farmers is often only $500 to $600 per farm year. This means that after deduction of the operating expenses as a rule there remains only an amount which is entirely insufficient to maintain an American standard of living.[80]

---

[80] *The Condition of Agriculture in the United States*, p. 68.

CHAPTER IX

HUMAN ELEMENTS IN COTTON CULTURE

CASE STUDIES

THE HUMAN factors in cotton, it is to be feared, have been too often disposed of in easy generalizations. These generalizations are likely to fall into one or the other of two extreme attitudes. White croppers in particular and Negro farmers in general, so runs one commonly held view, are inherently inferior, and their status as growers of cotton is merely an evidence of that basic biological fact. Cotton is financed and grown, another view holds, under a system by which small groups of designing men exploit the masses of southern farmers, black and white. It will be realized that the position herein taken is one that refuses to be bound by blanket explanations. Geographic, historical, and economic factors serve to explain many aspects of cotton culture. But the variations both within the practices in the cotton system and also in the traits of the human factors in cotton are so great as to render unsatisfactory many of the explanations offered in terms of race, heredity, tenancy, or the crop lien system.

The interplay of southern farmers and their cotton under widely varying conditions can best be shown in case studies. Specialization and diversification in cotton culture, management, integrity, industry, thrift, tact, and their opposites are set forth in stories of successes and failures of the human factors in cotton. Cotton wage hands, croppers, share tenants, small owners, and large planters are presented in case studies. Such cases as are here offered rarely find their way into the texts on economic history. They represent the average and the type of the agricultural process; they go behind the statistics and show the cotton system as it works. To those unfamiliar with southern agricultural processes, they possess, it is submitted, interpretative as well as descriptive value. The picture of cotton growers as they remain in the situation in which they were born, climb the ladder of tenure, or fail, offers material valuable for interpretation of the human factors in cotton.

THE COTTON CROPPER

Since he requires no additional capital the plantation laborer may easily set up as a share cropper. He needs only to be willing to take a part of the risk of cotton production and to gain the confidence of the plantation owner.

J. T. Walker, a Negro, of the Marshall farm community in Macon County, Alabama, while working as a monthly wage hand wished to test his ability in operating a small farm. He secured a plot of ground from his employer, which he worked at odd times while keeping up his regular work as a monthly wage hand. He made a very good cotton and corn crop on his small farm. His experiment was so successful that he gained confidence in himself and the following year, contracted with the plantation owner to work on halves as a share cropper.[1]

---

[1] T. J. Edwards in *Southern Workman,* 1917, cited in *Journal American Statistical Association,* XIII, 67.

Mobility, low standards of living, attempts to rise in tenure, and reverses characterize the lives of many share croppers. The following is the case of a white cropper from Texas:

A share cropper began for himself 24 years ago in Tennessee, and farmed in that state ten years as a cropper, moving five times. He has moved eight times and has had four reverses in tenure during the 14 years he has been in the Texas Black Lands. He attained the share tenancy stage and remained in it for one year only.

Eleven people, including a married son and his wife, live in a house of four rooms. The cropper has lost an average of four dollars a year since he began farming 24 years ago.[2]

The following white cropper, age 53, from the Coastal Plains of Georgia tells his own story. In 1913 he had moved eleven times in twenty years and had lost on an average $9.09 each year he farmed. His change of farms was the result of lack of success with each landlord. Success has come to be reckoned in terms of "corn, bacon, and money."

I have been a share cropper and shared with different landlords. I have six children living with me, two others grown and left me; one a renter and one a share cropper.

The first landlord I was with two years and left him with a little corn and bacon. No money.

Was with second landlord two years, and left with 50 bushels of corn and 300 pounds of bacon. No money.

The third landlord I was with three years and left with 50 bushels of corn and 500 pounds of bacon. No money.

Fifth landlord had me three years. Left with 75 bushels of corn and 500 pounds of bacon. No money.

[2] Dept. of Agriculture Bulletin 1068, p. 52.

The sixth landlord I was with three years. I left him with no corn, no bacon, and in debt $20.00.

Was with the seventh landlord a year. Left with 100 bushels corn; no bacon; no money, but out of debt.

Was with eighth landlord one year and left with 150 bushels corn, 800 pounds of meat and no money.

The ninth landlord I was with two years. Left him with 125 bushels corn, 800 pounds meat and $50.00 in money.

The tenth landlord I was with one year. Left him with 75 bushels corn, 500 pounds bacon and no money.

Was with the eleventh landlord two years 1911 and 1912 and left him with 100 bushels corn, 500 pounds bacon, and in debt $130.00.

Have been able to give each child twelve months schooling so each can read and write. Can read and write myself some.[3]

A cropper's statement of earnings presents the economics of growing cotton without other crops to furnish food and feed supplies. The excessive amount of man labor required for returns on cotton is also evident:

I, James Smith tenant farmer for Charley Childers on what is known as the N. M. Dellinger farm located two miles West of Lincolnton, N. C., make the following statement of my cotton farming for the year of 1920.

The landlord, Mr. Childers furnished the stock, land and tools, while I did the labor and the cotton crop was divided half and half alike, I receiving one-half and the landlord the other half.

13 bales of cotton grown at 15¢ makes a total of $700.42, cotton seed at 30¢ per bushel makes a total of $39.00, Total value of crop $739.42.

---

[3] D. S. Mead, Autreyville, Georgia, in F. J. Bivens, *The Farmer's Political Economy*, pp. 18–19. Pamphlet.

Less guano used under cotton $198. $541.42; Less ginner's toll, $48.00, Total net value of cotton, $493.42.

James Smith the tenant received one half, or $246.71.

By counting the number of days labor consumed in making and gathering said cotton crop, I hereby certify that my one half interest in cotton paid only 29¢ per day for labor producing it.

I am a cotton farmer and have been such for 17 years. In 1920 I planted only one acre of corn and no wheat, but placed my entire time and that of my family in producing the above cotton crop.

Dated at Lincolnton, N. C., this the 8th day of Dec, 1920.

JAMES SMITH.[4]

The restless, roving cropper continually looking for something better comes to be regarded as shiftless and dishonest. His mobility brings him into conflict with the landlord:

How may the unscrupulous tenant farmer be controlled?

About December 15, 1924, a large family of darkeys moved into a farmer's tenant house.

They got Christmas supplies, wood, house rent, etc., amounting to about fifty dollars.

In January, 1925, they moved out without notice for parts unknown.

About February 1925 a white family moved in and stayed three months. After getting house rent, wood and supplies for three months, they moved to a cotton mill.

Another white family moved in and stayed one month.

What is to be done with these dishonest, unprincipled, nomadic tenants?

There should be made a record of all of the sorry ingrates with the name of the landlord involved, the date of the of-

[4] Letter to the *Lincoln News*, Dec. 10, 1920.

fense, etc. All this should be presented in the name of a blacklist and sent to all farmers within one hundred miles.

Every tenant who applies for a farm should be required to give testimonials from former landlords as to character, dependability and industry.

This would make a better class of tenants and would protect the landowner against dishonest and sorry tenants.[5]

THE COTTON SHARE TENANT

The step in the tenure ladder from cropper to share or cash tenant is more difficult. To secure mules and implements requires industry, economy, and good crops. The Negro cropper seeking to rise in tenure status may meet with the reluctance of the landlord to relax supervision over his labors. The following case comes from Brooks County, Georgia:

Anthony More, a splendid worker, had a farm on this plantation, with 20 acres in cotton and 15 in corn in 1905. He made 13 bales of cotton and a large quantity of corn working as a cropper. The original account shows the large amount of cash this Negro got during the year, as well as other advances, and the credits of cotton, half the bale being credited in each case. He settled up the entire account and had corn and meat extra, not to mention the $100 in cash which he got just prior to the final settlement. This Negro abandoned the place the following year, because although he had enough to buy a mule the landlord would not rent to him for cash. He rented a farm from Mr. J. L. Brinkley, another Brooks County planter, and wound up his first year $70 in debt and nothing to show for the year's work. His former employer was interested to see how the case would turn out, as this was a good Negro so far as application to

[5] Letter from a landlord to *Monroe Inquirer*, Aug. 12, 1926.

work was concerned. The next year the deficit was taken up by another planter named Griffin. At the end of that year he was worse off than at the end of the former year. The Negro came to Mr. L. O. Borden, the original employer, last Sunday and told him of his troubles. Mr. Borden expects him to go back to work as a cropper, but has no doubt that as soon as he pays out of debt and saves enough to buy an animal he will again resort to renting. His observation is that after an unusually good crop year, it is very difficult to get labor for the following year. The Negroes have all made money and do not want to work again as croppers.[6]

A typical case of a tenant farmer using ineffectual methods of cultivation sanctioned by folklore comes from Tennessee:

An old Negro farmer, John Blake, renting 20 acres of good second bottom land, has lived with his present landlord 20 years. He now owns two "plug" mules, two cows, three hogs, one calf, a dozen or so chickens, a wagon and a few plows and tools. He owes the landlord $400.00. He has no fence around his yard or garden. He builds a split paling fence in the spring and uses it for kindling fires in the winter. He does not begin to clear his land of stalks or to plow and prepare it until April 1st, although our average planting date here is Apr. 20th to May 5th.

He plants his cotton, corn, and garden at certain periods of the moon. He is a close observer of the signs of the Zodiac, and their symbols all have a superstitious meaning to him. He would not violate these if he had to go hungry as a penalty for so doing. Thus it follows that his crops are planted on a hastily prepared seed bed, and through pro-

---

[6] Quoted in R. P. Brooks, *The Agrarian Revolution in Georgia 1865–1912*, p. 61.

crastination on his part they become stunted and choked with grass, and as a result he gets a low yield which sometimes pays his rent and living expense. Often it doesn't pay the rent, especially when price is low. Some years he clears a few dollars on his crop and immediately invests it in a Victrola, organ, family photo enlargements, or an automobile that has been driven at least 30,000 miles. You can guess the result of the story. The landlord usually loses on this class of tenant, and they are in the majority in this section. However, there are a few of this type of tenant who have taken more care and interest in saving, who are in good shape relatively and still rent land rather than buy it. But the majority of the Negro tenants as well as white tenants in this section are now in debt and in fact are seldom out of debt.[7]

Among complicating factors is the question of the inferiority of the tenants themselves. Isolated, ill-nourished, poorly instructed, badly housed, the tenant class, it has been suggested, is rapidly assuming the attitude of a dependent class. The presence or absence of the qualities of industry, good management, and thrift; the vicissitudes of nature and the market are shown in interaction with the cotton system in one case each from the Eastern and Western Cotton Belts.

The following case[8] of the struggle of a share renter in the plantation economy of the soil-exhausted Eastern Belt begins with the story of his father:

[7] Letter from a farmer, graduate of University of Tennessee Agricultural College, La Grange, Tenn., April 28, 1928.

[8] Adapted from the *Raleigh News and Observer*, Sept. 25, 1921. This represents an unusual type of feature story. Ben Dixon MacNeill went out a few miles from Raleigh, N. C., and interviewed an "ordinary poor white tenant." As written up it comes dangerously near the real case study method. It makes a vivid presentation of one human factor in cotton.

John Smith is as handy a name as any for this man whose house can be seen from Raleigh, whose story is set down here just as he told it not three afternoons ago. His father may be called Jim Smith.

Jim Smith was a day laborer, working on a farm in the eastern end of Wake County. He got eight dollars a month 50 years ago, and rations. By rations, his employer meant five pounds of meat and a peck of meal weekly. He plowed throughout the crop-growing months, and was paid his eight dollars. The agreement held until September, and then Jim Smith picked cotton. He got 35 cents per hundred, and his weekly wage could be raised to as much as five or six dollars.

The land was owned by Southerland Stewart. He had a farm of 600 acres, 18 mules, and two horses that were used for the family buggy through the week, and to hitch to the carriage on Sunday. He raised cotton, ginned it at his own gin, and in due season marketed it. About half the plantation population was the wage-earning negro, and half the wage-earning white men.

Jim Smith had married a wife. She couldn't read or write, but she could work. They both worked hard, she at a fixed wage per day, usually 25 cents, and Jim got his eight dollars per month. She hoed cotton all day, rising before dawn to cook breakfast and often cooking the three meals at night to do them the next day. The house was small, two or three rooms, and its furniture exceedingly scant.

Children were born, and for a few weeks the wife stayed home from the fields. Cotton picking time saw her again in the field, with the child left on a quilt in the shade of a bush nearby. Its mother suckled it occasionally, and picked cotton. The father picked cotton also at that season. In the spring the mother returned to the field, working if possible near her own house, and leaving the child at home.

Saturday afternoons the mother did the week's washing for the family and the father went off somewhere. Sometimes to a town nearby, and perhaps brought home a sack of candy to the wife and the children. If the mother had time after she had done the week's washing, she hoed the family collard patch. In the fall, the father went hunting, most likely, and Sunday for dinner there was squirrel.

Nine children were born to Jim Smith and his wife. They were reared in the field where the mother was at work, and by the time they were seven or eight years old, the smaller ones were left to care for themselves, and the older went to work. They were very handy at cotton picking time, and often the family income was swelled to the point that enabled them all to have a pair of shoes, or to go to the circus or something. Maybe the woman got a dress and hat that she could go to church in. They rarely went to church. Such social life as they had they got from visiting around the plantation on Sunday.

The children were not healthy. Fat meat and corn bread, with an occasional ration of biscuit are not conducive to health, nor is work at the age of six years conducive to strong growing. When John Smith was 23 years old, he married. The other eight children had died. Presently his father died, and his mother. John Smith was alone in the world, with his young wife. Neither had ever been to school a day in their lives. They had no vices, other than the use of snuff. They knew how to work.

They went to work. The first year as wage-earners, and the second as croppers. A new system had taken hold, and was rooting out the wageworker. The second year he bargained with Southerland Stewart for a one-horse crop, agreeing to plant 20 acres of cotton, and five acres of corn. That ratio was maintained all over the plantation. Smith was to get one-third of the cotton, one-third of the corn, and Stewart all the cotton seed.

Back of the big house where Stewart lived he had built a place that was called the commissary. Here the landowner sold to his tenants at prodigious prices, meal and meat, and flour and sugar and molasses, the simpler working garments, cheap cloth and shoes. John Smith plowed his crop and his wife hoed it. Together they raised a fairly good crop, and marketed it. That was 26 years ago.

They made 18 bales of cotton, for it was a good year and the land was good. It was the fall of 1895, and the country was recovering a little from the panic of '93. Meat had been costing four cents a pound, in regular stores, and Stewart had been selling it for no more than 50 per cent profit. John Smith knew nothing of these things. He just went to the commissary Saturday afternoon and got what they needed to keep them during the week.

At accounting time Smith had six bales of cotton up at the gin. Cotton was selling for six cents a pound. He didn't know how to figure it out, of course, but he felt very comfortable over his prospect. Also he had 25 bushels of corn as his share. He would buy him a hog next year. His cotton brought $206 and he owed the commissary just a little less than $200. He was disappointed, but not grievously. He had never had anything.

The next spring his wife had a baby. She had been working hard in the field all spring, and the baby died. She hadn't a doctor. An old negro midwife attended her. She was unable to do a full year's work, and the crop suffered. That year they made only five bales of cotton, and with the other expenses, they came out behind the game. They owed the commissary more than their cotton amounted to. They had their corn, and the hog weighed more than they expected.

Thirty dollars in debt to start with. Smith got discouraged. He thought he could do better somewhere else. Another landowner three miles away wanted a tenant, and Smith moved down there. The debt moved with him, his new landlord paying Stewart the $30, and adding it to the account that Smith started at the commissary. That year things were a little better, and Smith had $45 paid to him in actual cash money. He and his wife went to town and bought a bedstead and a cook stove, a hat for herself and a pair of shoes for him.

That winter they had another baby. Smith took a notion that he could make more if he had a mule of his own and got half the crop instead of a third of it. His new landlord sold him a mule on credit. At the end of the year he had $75. The baby had hindered the mother from work, and Smith had to hire some done. He paid $50 on the mule, and hoped that the next year he could pay the other $50. Mules were cheaper then. He bought a one-horse wagon, adding $30 to his debt.

It was a bad year, and Smith moved again, taking his debt with him to a third landlord. He kept his mule. That year another baby, and the $100 debt raised to $125. He got half the crop, but there was no pig. He didn't raise corn enough the year before, and he had to buy from the landlord to feed his mule. He wanted to plant more corn, but the landlord already had more corn than he knew what to do with. He couldn't sell it, and he could sell cotton.

Smith moved along to another place, moving his debt with him. Then struck a good year. Cotton moved up in price, and he had full ten bales, worth $500. He paid his debts and had a little besides. He bought some furniture. The next year he got in debt again and moved. The next he cleared out to a new location. Another baby that died. The first-born was big enough to pick a little cotton that fall. The mother kept the children in the field with her as she worked. He about broke even.

On this place there was no commissary. The owner of the land "stood" for Smith at a general merchandisery at a town nearby, who was to be paid first thing in the fall out of the proceeds of the crop.

Next year the boy would be ten years old, and big enough to plow. Smith bought another mule at the livery stable in town, giving a mortgage both on the one he had and the one he bought. The boy learned to plow that summer, and the family cultivated nearly thirty acres of land in cotton.

They raised 20 bales, and had to hire some of it picked out. They made nothing, but were able to pay some on the mule, settle up the store account, and buy a few clothes.

More children, more moving, more debts at the store, and a mortgage on the mules, the wagon, and about everything the man had. Smith took a notion that he ought to have a piece of land. He bought a little place, five acres, with a sort of a house on it, and moved in, paying part down. He had his mules and his boy that could plow and another that could almost reach the plow handles, and a wife with still a lot of work in her. The place was too small and he let the mortgage take it and moved on to another farm the next year.

Disaster overtook him that year. Crops failed, prices failed, the new landlord lived in town and didn't care about John Smith. He sold him out, took his mules, his corn, his all. John Smith was back where he started, with four children, two beds, a few other pieces of furniture and two dogs. He went back on a third-crop basis, and began again the hard, hopeless task of getting himself on a paying basis. History repeated its grim cycle, more work, more children, more debt, more moving from place to place.

And now the John Smith today. Of the world's goods he has: three mules, two pigs, 11 chickens, a wagon, four plows, harness, an old buggy, a few squalid pieces of furniture, a wife, seven children, two cur dogs, and is in debt $1,400. This year he has spent at the store $250. He will make 12 bales of cotton, half of which he will get, and turn over to the man to whom he owes the money—the storekeeper, etc.

Three prosperous years are immediately behind him, in the common estimate of the conditions of the farmer. He has not moved in that time. He was $600 in debt three years ago. Half his new indebtedness is chargeable to an epidemic of typhoid fever that struck down six members of the family at one time in 1919. The entire working force of the family incapacitated for a summer. The winter before that all of them had had the influenza.

John Smith made one desperate effort to pull out winter before last. He succumbed to the popular frenzy of violation of the Volstead Act. He became associated with one of his neighbors in the manufacture of liquor. It promised dividends, liquor was selling for $12 a gallon, a bushel of corn would make several gallons. Profits would be large, the work easy, the risk nominal, and the initial investment small. His boys helped, and the partner attended the business of marketing.

Again disaster befell them. Raiders got the rude, homemade plant one day, and destroyed the collected raw material. John Smith retired from the liquor business. He had failed at that, as at everything else to which he had put his hand. It was the only time that he had ever broken the law. He escaped uncaught, to be sure, but the loss but added to the already mountain-high debt. He drank a little of his product, but not much. He has always been a sober, industrious tenant.

A brief analysis of the record of John Smith, Tenant Farmer, follows:

He has in 28 years, with the help of his family of seven children, produced 336 bales of cotton that brought $16,600.

He received for his share of that amount $7,900.

He has received an average yearly compensation of $282.15.

Out of this he has lived, clothed his family, bought six mules and partly paid for them.

He has spent on an average of $328.56 annually over a period of 28 years.

He is now in debt $1,400, and when his mules were taken from him ten years ago, his creditor lost $600.

His present liabilities arc $1,400, and his assets are $700.

The expenditure of $328.56 includes such permanent improvements as the purchase of mules, wagons, furniture, etc. It also includes money paid for food, clothing, doctors and medicine, etc.

Added to these, of course, must be some small allowance for vegetable gardens, corn and potatoes, and an occasional hog and chicken that has been raised on the place. Such things are generally discouraged by landlords, since they are not directly productive of returns to him. Ordinarily the tenant lives out of a store for ten months in the year. He pays in the fall.

There are seven children living and three have died. They all work in the field and none of them have ever been to school a day. Although underweight and undernourished they appear intelligent and reasonably healthy.

A case from the Texas cotton area shows large acreage planted, greater production, a smaller amount of the family living secured from the farm, and equal difficulty in buying land:

L. T. Steward,[9] a tenant farmer of Savoy, Texas, described before the Federal Commission of Inquiry into Industrial Relations, at Dallas, Texas, his efforts for twenty years to buy a farm. He began in Faulkner County, Arkansas, and finally came to Texas. After his first year's farming he sold his mule to "get square of debt." Next year he borrowed a mule and "came out $15 to the good." Then he bought a small farm on six years' time, but was forced to give it up for lack of $40 to meet payments at the end of the first year. After several years he got $200 ahead and bought an 85 acre farm in Arkansas. He did well but two children died and their doctor's bill cut into his savings so that he gave up the farm unable to meet the interest. He did not live more than two years on the same farm after he married. Some of the years he worked land controlled by relatives.

---

[9] Summarized from his testimony, in *The Land Question in the Southwest,* Report of Commission on Industrial Relations, IX, 9006–44.

He raised from four to eight bales of cotton from an average of thirty acres each year. Usually he ended the farming season with nothing.

Thinking Texas offered better opportunities for the tenant farmer and believing health conditions better, he decided to move west in 1903. The history of his farming follows:

Rented 50 acres of black land in Lamar County. Moved after the first year. Rented 45 acres of sandy land. Moved at the end of the year. Broke even both years; made 11 bales of cotton and 1,000 bushels of corn. Went in debt at the beginning of the next year. Cut wood at times to increase earnings. Lived on this farm two years, making $200. Wet land prevented good crops. Moved south of Paris in 1907. Cultivated 55 acres, five of which were in corn. Incurred doctor's bills and gave notes for store accounts. Moved at the end of the first year. Made $200.

He did not stay on this farm, he said, because the house was not good, no schools and unsuitable surroundings. Lived two years 13 miles from Paris. Worked 35 acres of corn and 75 acres of cotton, making 34 bales of cotton and 750 bushels of corn. No more than broke even the first year because of high prices of commodities and low price of cotton. Gave mortgage for provisions. Raised fair crop next year. Sold cotton to buyers who told him the grade. Steward does not know anything about grading cotton.

Moved three miles away the following year and cultivated 125 acres. This year he bought the implements and team of the farmer who left the place Steward rented. He then began as a "third-and-fourth" farmer. Made 48 bales of cotton and 100 bushels of corn. Paid for his pair of mules and wagon and tools and had 800 bushels of corn worth 65¢ a bushel, but no money. He moved at the end of this year because the landlord sold out.

In another part of the county he farmed 120 acres. Made 52 bales and 800 bushels of corn the first year. Cotton sold for 12¢ to 15¢. Broke even. Store account for this year totaled

$1,700. Said he did not keep check on what he purchased. Did not know how much he owed till time came for him to pay up. Groceries consisted of meal, flour, meat, some canned goods, syrup. Lived very well but not elegantly. Did not dress any better than when living in Arkansas. Clothing bill for the entire family at that time amounted to about $75 a year. Ten-dollar suit lasted him five years. Usually wore overalls and duck trousers.

Second year on this farm he made 57 bales. Store account totaled $1,250. Cotton cheaper. Made 87 bales the third year, selling at from 6¢ to 11¢. Store account $750. At the end of the three years worse off than at the beginning. In 1913 he worked 90 acres for the same man and made only six bales of cotton. Finished $700 in debt. Secured debt with his live stock and implements. Moved to Mulberry Bottom. Made 27 bales from 100 acres. Went deeper in debt. Paid for corn and cotton picking, but still owes $600 grocery bill. Moved to Fannin County. Living today in a house which he has rented for two weeks. Has no prospect for getting a farm this year. He has lived in not less than 20 different houses, produced at least 450 bales of cotton, several thousand bushels of corn, made two attempts to purchase a home, is the father of eight living children, is in debt $700 and practically is without means of sustenance.

School advantages for the children have not been good. His oldest son spent parts of three years in school. At times he had lived too far from the schoolhouse for his oldest boy to attend. Then when he lived near town later on the boy was ashamed to go on account of his clothing. His oldest son started work on the farm at the age of seven. That is the "corn dropping" age, he said. The next step in farm employment for the child, he told the commission, was to chop cotton.

"Always I wanted to do as best I could," Mr. Steward said. "I never had a roaming disposition. I was dissatisfied because I had little to say about what I planted. My observation is that tenant farmers do not stay in the same place very long. If the tenant farmer gets to making money, he's got to move. If the landlord don't get all the money, he wants the tenant to go farther."

Mr. Steward said he is a member of the Woodmen and Odd Fellows' Lodges. He said he had voted three times in his life, and once in Texas. Asked why he did not vote, after he had said he always paid his poll tax, Mr. Steward said "it didn't do any good, for things went the other way."

His view about renting on the "third-and-fourth" plan is that a person having good land can make a living. He believes landlords should better the rent houses. He said he thought much of the sickness in his family was caused by poor housing conditions.

Chairman Walsh questioned the witness about his understanding of politics. Mr. Steward said he understood the Ferguson land plank and had read something of the issues advocated by the two old national parties. He said landlords did not as a rule attempt to influence the votes of tenants, although he had heard some landlords say they would not rent to Socialists. He has studied some and his current reading has been confined to the *Appeal to Reason*, the *Buzz Saw* and the *Fort Worth Record*.

He told the commissioners that he had interested himself in modern farm developments to a certain extent, but had not accomplished very much.

"The Texas Industrial Congress sent me a letter of recommendation and an emblem for being a good farmer," he said.

Mrs. Beulah Steward gave a brief statement of her life on the farm. She was married at the age of 15. Their household furniture was bought on credit. She estimated that she and her husband had lived in at least 20 different houses. These ranged in sizes from one to seven rooms. She worked practically every year in the field, starting by daylight and quitting at sundown during the farming season. The family ate three meals a day. Breakfast usually was eaten at 4 o'clock in the morning. With the aid of a sewing machine, which was taken from her last January, she earned as much as $2 a day some days by sewing for neighbors. She had to work late into the night to do this. She has not worn a ready-made garment nor read many newspapers since marrying. Sometimes she read serial stories in magazines. She has owned three hats during the last 25 years. Sometimes she "got to town" only once in two years. To make the best of her condition and not worry about anything, she said, is her motto.

The restricted standard of living of the cotton tenant may find expression in times of rising cotton prices in unwise expenditures. Pent up desires, reënforced by a belief that the next cotton crop will bring as much as the last, lead to the purchase of luxuries that are ill-afforded. In the case quoted the farmer suffered the tenant's greatest tragedy—the loss of his work stock. It is worthy of note that one of the commonly accepted attitudes in the South is to lay the ills of the cotton farmer to the purchases of automobiles in times of prosperity:

A very unpleasant duty of an Iredell county deputy sheriff was performed last Friday, when James F. Aldmon received seizure papers from Cabarrus county in which a former citizen of Cabarrus county had mortgaged two mules and two cows on security for a secondhand automobile to a Concord dealer. The tenant was coming in from the field with his mules at the noon hour and when they were divested of their gears, the officer laid claim. One of the mortgaged cows had died since the papers were given in exchange for the machine. The tenant was left stranded as to mules for the working of his crops and the cow which furnished milk and butter for the family was taken. The scene was almost tragic, and the heart of the officer melted away when the woman of the household prayed for the deliverance of the ones holding the mortgage from a life of torment in the next world. The live stock was taken to a neighbor's house, to be held for a few days, giving the tenant time in which he would be permitted to raise the amount due on the mortgage.[10]

It is possible for cotton lands to be rented as a capitalistic undertaking. In the case here cited, cotton tenancy corresponds more to the form of tenancy in the wheat belt than in the South. This case comes from the Mississippi Delta of Arkansas:

Chalmers Barnett

Married spring of '24, $300.00 worse than broke. Rented average land. Bought team, tools, feed, etc., on credit. Made good crop. Got poor price and paid all.

1925. Bought another team and rented 40 acres more land. Better tools and a Ford car. Good crop. Poor price but paid out.

1926. Rented 80 acres more land. Bought two more teams. good crop. Got poor price and paid all.

1927. Worked a 5 team crop (160 acres). Poor crop. Good price. Paid out.

1928. Rented 40 acres more land and bought another team on a credit prospect of a fair crop and a moderate price, so should pay out.

This man works well and is far above the average in practical ability and intelligence, and I am sure will succeed and go further. He grows cotton exclusively. Buys all feed at large price.[11]

10 *Greensboro News*, (N. C.), June 26, 1926.

11 Letter to writer, March, 1928.

THE FARM OWNER

The rise from share tenancy to land ownership is best accomplished where industry and application on the part of the tenant is met by an attitude of helpfulness on the part of the landlord. This attitude often exists in instances of cordial personal relations between tenants and landlords:

> Now I am a plain farmer and own a four-horse farm myself and work three families of tenants on my place, and I had rather help each one of my tenants to own that part of the farm that he works than to have to use them as tenants all of my life . . . if there was any way opened up to them for the help that would enable them to purchase.[12]

The way to land ownership in the Cotton Belt for the tenant is hard and difficult. If landlords are coöperative, the rise to ownership may be blocked by vicissitudes of fortune or the lack of indefinable but necessary personal qualities.

> The writer, a landlord in Upson County, Georgia, soon after the Negroes were freed undertook to aid many of his old family Negroes with disastrous results in all except one case. Will mention one case of late occurrence to illustrate. Fourteen years ago I bought a farm of 120 acres and a fine mule and wagon and other items of supplies for one of my favorite Negroes, and sold them to him for what they all cost me. During that long time he has only paid the cost of the mule and the wagon and last year permitted the land to be sold for taxes.[13]

---

[12] Letter from F. A. Marsh, R. F. D. No. 8, Fayetteville, N. C., to *Raleigh News and Observer,* Oct. 1921.

[13] Given in R. P. Brooks, *The Agrarian Revolution in Georgia,* p. 61.

Racial inferiority, sometimes assigned by popular opinion, is no blanket cause for failure to rise in the tenure ladder. The story of the following Negro tenant shows the alterations of crop yields and price. His expenditures in years of prosperity indicate a desire for a higher standard of living:

> The following is the history of Harrison White, a renter of Tysonville, Macon County, Alabama, and his ten year struggle to get money to start buying land. In 1901 he farmed with one plow, paid himself out of debt, and cleared $200; in 1902 he continued farming with one plow and came out $175 behind; in 1903 he came out $388 behind; in 1904 he ran one plow, made 13½ bales of cotton, paid up back debts and made 250 bushels of corn; in 1905, to make a crop he bought an ox to supplement his over-worked horse. He paid his debts and cleared $380. He paid $35 for a wagon, $60 for a buggy, $55 for a sewing machine and $45 for furniture; in 1906 he bought a mule on time for $225 and came out $190 behind; in 1907 he paid off $190 with interest, bought another mule and saved $200; in 1908 he saved 3 bales of cotton and much corn; in 1909 he paid out and cleared one bale of cotton; and in 1910 he paid all debts and banked $300 to invest in buying land.[14]

When needed personal qualities in tenants meet with integrity, fair dealing, and coöperation from landlords, the rise to ownership may be rapid. In the following case social advancement accompanied economic advancement:

> Possibly it will be helpful to relate what was done for three tenants down in Franklin County, whose cases came under my personal observation. These three "Smith Boys,"

[14] J. T. Edwards in *Southern Workman* 1917, cited in *Journal American Statistical Association,* XIII, 68–9.

as I shall call them, rented a tract of land containing one hundred and fifty acres. The owner of the land became interested in them and told them that he would like to see them own the land they tilled. They replied that it would be foolish for them to undertake to buy the land as it was all they could do to pay the rent. The owner said:

"I am rather surprised that you do pay it. You are not half farming the land and you are not to blame. Why should you make my land rich? Next year I may get another tenant or may sell the land, and your labors would be lost. If you buy the land, you can afford to bring it up and can more than double its producing power."

The Smith boys were timid; they had no faith in themselves and doubted the good faith of their landlord. But he insisted and practically forced them to buy the land on eight years' time. Soon a change commenced to come over the face of the land. One saw ditches and fence corners cleared up, terraces run here and there. The crops were rotating, orchards sprang up, outhouses were white-washed, and in five years one looked on a new land.

The change in the Smith boys was greater than the change in the land. When they came to town, there was a spring in their steps and a light in their eyes never seen before. They worked hard all the week, and often the women worked with them. But on Sunday, when they put on their best and went out to church, they looked like "Lords of the Soil."

They met every payment except one and that year there was a total crop failure on account of the drought, and the man who sold them the land simply said, "Forget this year's payment and make it nine years instead of eight." I will make no comments upon this story, but let the facts speak for themselves.[15]

---

[15] Former Governor T. W. Bickett, North Carolina, in *Raleigh News and Observer*, Oct. 2, 1921.

Southern farmers, without training in the schools, have risen by virtue of sheer physical labor to landownership. The lack of education is, however, likely to be reflected in poor cultural standards and inadequate care of the soil. The following case comes from Tennessee:

Mr. Watkins, twenty years ago, began renting land and growing cotton. He is one of the hardest of workers, in a physical sense, but illiterate. He lives at home; buying little or none of meat, lard, corn, preserves, canned goods, vegetables, fruits, chickens, eggs, beef, etc. He raises all this at home. He wears denims and so does his family. All the family worked at the home and in the fields. Altho unscientific in his methods he had lots of good conservative ideas about farming and crops. He always has grown plenty of feed crops for the farm work stock. He cannot read or write, but hard common sense earned in the school of experience is his chief asset in business methods. Financially, he was poor, but always a shrewd but honest trader. His earnings he invested in small tracts of land until now he owns about 500 acres with a comfortable home, gin, saw mill and grist mill on it. He buys, sells and trades livestock the entire year in addition to farming. He has been very liberal with those who helped him make and save. He has given each son and daughter 50 acres to begin on as they marry.

The worst of this story is that Mr. Watkins is a *soil miner;* taking off crop after crop and putting back nothing, and his land is showing it. It also is washing badly due to lack of terracing and proper drainage. Land is cheap here, but is broken and somewhat eroded. About $15.00 to $20.00 per acre is the prevailing price of farm land. Approximatcly ⅓ to ¼ tillable land is the average. It produces from ¼ to 1 bale per acre, depending on fertilization and methods used. But it takes a real manager to make a profit at it. Mr. Watkins needs the help of the County Agent badly and will not take advantage of it. He made what he has with hard work, living at home, and saving by following a standard of living that most informed people are not willing to follow. He does not use fertilizer but does plant legumes on part of his land every year. He has learned only one step, in my opinion, in the true economic and scientific production of cotton on a profitable basis.[16]

A hard-working tenant may get ahead enough to buy a farm with the aid of a loan from a Federal Land Bank. The following is the case of a farmer in the Arkansas Delta:

Harry Barnett

Married in August, 1921. Rented 70 acres of cotton land in 1922. Had everything to buy. Owed $400.00 to start. Paid out and had a bit left. 1923 about same, and paid a few hundred on 80 acres of land.

1924 rented 40 acres more land and another team and made small payment on own land.

1925 rented more land and bought two more teams.

1926 bought 200 acres of good land for $19,000.00, turning in equity in 80 acres at $1500.00. Bought more teams and worked same land as heretofore, besides 120 acres of own 200 acres. (C. Barnett rented 80 acres of his land.) Rent of farm paid all interest and $1000.00 on principal.

Bought 250 good steers ($15,000.00) and fed for ninety days on own land, thereby doubling production upon 60 acres, which was a good profit on cattle. Also made $700.00 additional profit on steers.

1927. Got $8000.00 Federal Land Bank Loan, upon 200 acres @ 5% int. and gave T. J. Williams second mortgage for $8300.00 @ 6%, being balance due him for purchase money—$1500.00 of which I've paid, besides all interest on

16 Told by a planter of La Grange, Tennessee.

both mortgages. Also paid some carry-over from 1926 for share croppers, and have $500.00 left for 1928. I now have a net worth of $8000.00 and my indebtedness on farm will be cared for from rents from same, besides giving me $1000.00 a year to apply upon principal.

I grow nothing except cotton, one-fourth of which goes for rental, which usually amounts to $10.00 to $15.00 per acre. So it pays to buy land with 5% money.[17]

The rise of an able cropper, the fall of his fortunes, caught in the past war deflation in cotton and land values, and his rise, with the aid of a banker friend, in the Arizona Cotton Belt is set forth in the following case study:

M. O. Parker, Phoenix, Arizona,
the Best Farmer I Ever Knew

At the age of 30 had accumulated $5000.00 farming in Missouri, paying one-half of all his production as rent. Moved to Louisiana in 1914 and lost out in two years. Under a partnership deal with the writer in the sale of some plantations he received clear profit in 1917 and 1918, one-half or $10,000.00 cash at one time. Bought 160 acres with all stock and equipment in early spring of 1920 near Rogers, Ark. Worked hard and lived close for three years. Went in with $11,000.00. Came out with $700.00 and three years older. Pretty badly broken in spirit as well as in cash.

I gave him free of rent for two years 40 acres of rich land adjacent to Phoenix, except for taxes about $9.00 per acre. Loaned him $1400.00 1st year and I think $400.00 second year. In the meantime he rented other lands up to 200 acres. At the end of three years he bought this 40 acres of me for $19,000.00 net, paying me $4000.00 cash. In 1926 he paid all due $900.00 net and $500.00 principal. In 1927 he paid $4000.00, although only $1360 due, and is now easily

[17] Letter to writer, March, 1928.

worth $20,000.00 and going strong, specializing in long staple cotton.[18]

The death of work stock often causes a reduction in the tenure of the cotton farmer. The following case [19] shows the relation of a small owner in the Coastal Plains of Georgia to the credit system:

Runs two plows; works children in field as laborers—girls and boys.

"My horse died and I had to get another horse on credit to work the land with. Made good crops, lived economically, plain clothing and no luxuries—doing without many necessities. In fall 1912 my creditors did not give me time to gather all my cotton to pay as far as I could, but the remainder of my cotton was levied on in the field, and also my corn and potatoes were levied on and sold by the sheriff. Having lost all my eatables, the purchaser felt sorry and gave me back the sweet potatoes."

Three cases of cotton farmers came from the Coastal Plains of Georgia. All are fairly well-to-do and above the average small owners. The first attempts diversified farming, the second has lived off the sales of timber, the third has a long-time farm loan:

1. Owns 200 acres of land, paid for and owes no money, and does not borrow or buy on credit. Runs three plows; has eight children. Hires no labor, but works children as laborers in fields. Dresses himself and children in only ordinary common clothing, and has the necessities of life. Makes an average of twenty-five bales of cotton a year.

"I live at home; I buy clothing, shoes, flour, coffee, sugar and fertilizer for which I pay cash. I am a diversified farmer

---

[18] Letters from an Arkansas Banker, March, 1921.

[19] Bivens, *The Farmer's Political Economy,* pp. 18–19.

as taught by the newspapers. I have not had any providential misfortune. I want to make some money and have tried, but cannot do so growing cotton. I am sore on cotton farming, as it is impossible to be a thrifty citizen and grow cotton. There is no wage in it. You cannot grow cotton to profit by hiring your labor, and where a man does not work his children from six years of age up in the field, he cannot come out of debt and also have the necessities of life. He could not do it if his life depended upon it."[20]

2. Runs two plows. Works only himself and children. Is a diversified farmer, growing all his eatables at home except flour and sugar. His labor is done by himself and three sons. One son is of age and will farm for himself during this year, 1913.

"I give my average income and expense for ten years and show how I stand:

"My expenses for a family of seven: clothing, $100.00. Flour, sugar, coffee and other necessities, $100.00. Swapped old stock for younger average per year, $30.00. Improvements, building, fences, etc., $100.00. For church and school purposes, $70.00. Insurance Premiums $63.62. Fertilizer $100. Labor $75.00. New Tools average $50.00.

"Make from 8 to 14 bags of cotton and handle from $500.00 to $700.00 per year. Average outlay above $738.62. I keep books. Here they are.

"You ask how can one pay out more than was taken in. I will explain: I sold my timber to make the improvements with the $100.00 item mentioned and this winter borrowed from N. Y. Life Insurance Co., $125.00 on my policy. I paid my twenty-one-year-old boy $75.00 and my bank account is now $10. You see in the past, I had money in the bank at times of from $100.00 to $500.00 from the sale of my timber. That is spent living at home economically, pro-

---

[20] Mr. Alexander Hull, Moultrie, cited in *The Farmer's Political Economy*.

viding ordinary clothing and doing without what was wanted except necessary things; have not kept even growing cotton. People can talk and theorize, but I have been and am going to mill. I owe no money and have no interest to pay. I can borrow money from the bank or buy on credit. I have grown cotton in the past to buy those things that I do not grow or make, but after ten years of good management I am now compelled to grow cotton to pay my running expenses. I run behind to the tune of about 2 cents a pound on my cotton.

"We farmers will have to quit buying, or debts will get us and ours. What makes Georgia seem to be prosperous is that millions are loaned each year from long loan agencies for periods of five to ten years. If I had a long loan interest to pay I would have to give up right now. I could only afford jeans for clothing and work barefooted. The long loans are procured by farmers to square up with. There is no money in growing cotton at 15 cents a pound and even at that price a farmer could only get the necessities of life, at a less cost he cannot. I would say that I am above the average farmer." [21]

3. "I own my land, run four plows and work my children as laborers. We make an average of thirty bales of cotton a year. I borrowed from a long loan company to improve with. After paying taxes, fertilizer, clothing and shoes, I have no money left. When my five year loan fell due I had to borrow from another loan company to pay up with and had to borrow a little more this time to fill up the hole caused by trying to have part of the necessities of life. There is no money in growing cotton for me, but there is nothing else I can grow for market except stuff for a home market and this is limited. I have no money now and will have to borrow from the bank or get goods and supplies from a credit store." [22]

---

[21] A. A. Moshburn, Doerun, Ga., cited in *ibid.*

[22] A. P. Tucker, Moultrie, Ga. cited in *ibid.*

Set off in contrast are the cases of farmers who have prospered amid the fluctuations of cotton by growing a diversity of food and forage crops with poultry and live stock. The advantage of diversification is shown by the case of (1) a Negro owner of a small farm in Alabama, (2) two North Carolina small owners, (3) two contrasting cases from Arkansas, (4) two purchasers of cut-over timber lands from the same state, (5) cases from Mississippi of absentee and at home farming, (6) a large scale planter in Alabama, and (7) a scientific agriculturist in Mississippi.

1. John Griffin is a Negro farmer, who lives about four miles from Tuscaloosa, Alabama, about fifty years of age; and he is a man of good character and reputation, well liked and a good citizen. A few years ago John was a small cotton farmer and borrowed money from the bank to make his crop, —generally $400.00 to $500.00. He usually was able to pay up and borrow more the next year, but he was not getting ahead. He was advised to get some milch cows and sell milk. A few cows, milked by his wife who also marketed the milk, did not lessen his farming operations. The dairy business was a side line. These few cows in a short time paid for his farming operations and left him his crop free. He gradually increased his herd until now he has about seventy head, among them some registered cows and a registered bull.

At this time John Griffin owns about one hundred sixty (160) acres of good land, has money in the bank and loaned out.[23]

2. (A) A man who cannot read nor write, not even his name, bought a farm twenty years ago from a man who sold it because the farm was so poor that he could not make a living on it. The man who bought this place had worked

---

[23] Letter from an Alabama banker, March, 1928.

in cotton practically all of his life, but he stopped making cotton except about one bale per year and sometimes three or four bales, but he raises everything to eat. He carries a load of cut wood to town each time he makes a trip and thereby makes each trip count. He has pigs, a milk cow and some of almost everything that he raises for sale each year. He values time and he values money. His farm is now in good condition, equipped with mower, reaper, wagons, teams, etc. It is paid for and he has money out on interest. He has an automobile but he paid for this when he bought it, and now he practically pays for his fertilizer each year with the interest that he receives from the money that he has loaned.[24]

(B) A man who raises cotton but is never compelled to sell it on a low market. His wife sells chickens, strawberries, cantaloupes, vegetables of all kinds and they are willing to make sacrifices to keep out of debt. This man has an automobile, but it was not bought on a credit.

3. (A) One farmer raised cotton first and made an abundance every year and raised a little feed and practically no food. He had practiced this policy prior to 1912 and succeeded fairly well until the year 1912, and of course during the period of the war he made money, but the year 1920 practically broke him, and during the past six or eight years he continued this policy until he lost his farm, live stock and everything he had, his life time earnings.[25]

(B) Another farmer of about the same class so far as size of his farm and equipment and equally as industrious, farmed on the plan of raising his feed and food first and his cotton second; has not lost his farm. He apparently did not make as much money during the period of the war, but his losses in 1920 were not disastrous, and he has gradually

---

[24] Letter from a N. C. merchant, March, 1928.

[25] Letter from planter, Helena, Ark., April, 1928.

pulled out of the loss he sustained in 1920 until he now has his farm practically free of debt.

4. (A) The two cases which I will give you are called Mr. Brown and Mr. Jones, which are far from the names. About twelve years ago the company which I represent finished cutting the timber off of some of their holdings and built houses on it in the cut-over state and offered the farmers what they could make on it three years for putting it into cultivation. Mr. Brown moved on 40 acres of this land with a wife and five children, a pair of small mules, about 60 bushels of corn, a few tools, one milk cow, about 30 head of chickens and about a 40 pound shoat and $85.00 in money. Today Mr. Brown owns 110 acres of land on which he owes about $4,000.00. He owns 6 head of good mules, 5 good milk cows, about 150 pure bred chickens and 12 or 14 head of registered Poland China hogs, has a nice home orchard, has his place well fixed up with necessary out buildings and conveniences, owns his own automobile, pays cash for everything he buys, and borrows no money. He is a member of the school board in his local community and is an upstanding leading citizen.[26]

(B) Mr. Jones moved on just as good a piece of land on the same kind of trade about the same time with a wife and three children, 4 head of mules better than the mules Mr. Brown had, but no cows, no chickens, no hogs, a Buick automobile, and $600.00 in money. When we foreclosed on Mr. Jones two years ago because he was dropping further and further in debt we charged off to profit and loss a little better than $2,000.00, and today Mr. Jones is making a share crop. His son is making a share crop and both of his daughters have married share croppers. There was no difference in the two men's conditions, but there was a big difference in the personal equation. One man attempted from

[26] Letter from manager of a lumber company in Ark., March, 1928.

the beginning to raise as near as possible his feed and food supplies and have something to sell with which to pay for what he had to buy, planted the balance of his land in cotton, which belonged to him when it was picked; the other man did not believe that it paid to grow garden truck, corn, hay, pasture, and things like that on $150.00 an acre land that could make a bale of cotton per acre; therefore, he lived from the grocery store and depended on his cotton crop to furnish his living twelve months a year.

5. (A) First: We have a negro farmer who owns and operates 200 acres of land. He lives on the land. He has always grown his feed crops, seldom buying any feed at all, and produces all the cotton he can after his feed crops are provided for. He has never borrowed any money but has been able to maintain a splendid balance in the bank from year to year.[27]

(B) Second: We know a man in the same community who owns a tract of land consisting of more than 500 acres, who does not live on the land, but lives in town in a good home. He grows some of his feed, makes fairly good crops—in fact we would say about the average crop of both cotton and feed stuffs, but whose living expenses have always been regulated entirely out of proportion to his income, and as a result his 500 acres is heavily mortgaged, his home is heavily mortgaged and though a man of high degree of intelligence, of good habits, there is some doubt as to his ability to ever remove the mortgage from his property

6. As an example, we refer to one party in particular whom we shall refer to as Mr. Doe. This particular farmer began some 20 years ago with a limited capital of approximately $200.00 which he applied as part payment on 20 acres of land. From time to time he has increased his holdings but has always seen fit to raise his own feed stuffs. Most of his activity is confined to cotton planting and last year he

---

[27] Letter from banker, Shelby, Miss., 1928.

raised and ginned approximately 1500 bales of cotton. Incidentally, he owns approximately 2,000 acres of land in an adjoining county, and our information is that on this 2,000 acres he only owes approximately $3,000.[28]

7. Near my old home a farm was purchased some ten or twelve years ago by R. C. Small, who was at that time superintendent of the McCormick plant of the International Harvester Company. It was about as poor a farm as could be found in the county. It was lacerated from erosion. The first year he plowed the land and sowed it to oats, he hardly obtained more than five bushels per acre. He followed that with peas, which were turned under. Last year he gathered something like 80 bushels of oats per acre. The land planted to corn must have yielded 75 bushels per acre. It was not necessary for him to buy any commercial fertilizer. It was all grown right there on the ground. No more fertile farm can now be found anywhere in the county. He put on some dairy cattle, and is selling about $200 worth of dairy products per month. His farm is handled largely with a Farm-All Tractor. The Farm-All Tractor now will afford opportunities for the cultivation of considerably more land than has ever been cultivated heretofore with the same amount of labor, and the farms above mentioned could be easily doubled by the use of a Farm-All Tractor, though the Farm-All Tractor would be justified if only 20 acres were under intensive cultivation, and the other 60 put to other crops, oats and cover crops.[29]

An unusual cotton farmer was Sam McCall, age 75, an ex-slave of Alabama. He afforded a significant contrast to John Blake (*see* page 258). His method of diversification and rotation was made the subject of a bulletin[30] by the United States Department of Agriculture:

[28] Letter from an Alabama banker.

[29] Letter from executive of a lumber company, Miss.

[30] M. A. Crosby, *An Example of Intensive Farming in the Cotton Belt,* Farmers' Bulletin 519, 1913; Smith, *North America,* p. 260.

Sam McCall decided to remain as a tenant on the plantation of his former owner in Alabama. Being thrifty, a hard worker, and a good manager he saved what he earned and was soon able to make a payment on 40 acres of land. He thus started farming for himself, and later added to his purchase until he owned 163 acres. He became a one-horse farmer in a section where the land was so grown up with brush and cut with gullies that the exhausted soils had been abandoned by the so-called better class farmers.

After growing cotton and corn for 15 years by the accepted methods on as much of his land as was tillable, McCall decided to intensify and diversify his farming. He selected his two most fertile acres and worked out the following method of farming: Oats were grown, harvested, and the stubble plowed under. Immediately afterwards he planted corn, and in April he planted cotton between the corn rows. Cowpeas were then planted between the hills of corn. As soon as the ears of corn were established he cut off the tops above the ears to let light in on the cotton. As soon as the corn was matured he removed the stalks, thus giving the cotton full possession of the ground during the latter part of the summer. In one year one of these acres yielded 75 bushels of winter oats and three bales of cotton. The other acre yielded three crops, one bale of cotton, 50 bushels of corn, and 50 bushels of oats. He increased the fertility of the soil without the use of fertilizer by plowing under cowpeas, oat stubble, and corn stalks.

By selecting the seed from the best plants he developed a high yielding strain of cotton, known locally as the Sam McCall cotton which he sold at fancy prices. He soon found that his best plants were producing a pound or more of lint apiece, and he figured that with a perfect stand of plants he could produce nine bales of cotton on one acre. He has secured at one time as high as seven bales on the two acres. One year he succeeded in growing a 500 pound bale on a measured one-eighth acre which produced a perfect stand of

612 plants. Sam McCall, a naïve Negro farmer, had become something of a creative artist working with the soil.

For the last 15 years of his farming he received an annual money income of from $200 to $350 from two acres, more than his tenant neighbors who cultivated 20 acres. With his living from the farm, he had, J. Russell Smith suggests, a better income than the average farm hand or city worker of the same territory and time, and his work left him in better condition and with more leisure time than the average factory worker.

The effect of the boll weevil invasion on the effort of a southern farmer to pay for his land is recounted in a case from South Carolina:

Herbert Rawlings, my father, bought a forty-two acre farm in southern South Carolina for $5,000, paying $2,000 in cash with the balance in yearly payments. Of the six children the four eldest boys were able to work on the farm. The family improved the place the first year, raised some hogs, kept a cow, and grew food and feedstuff. But we depended on cotton to pay for the farm. That year we made a big cotton crop, but a storm came and blew bales of cotton out of the bolls on the ground. Cotton brought a low price that year, and we did not make very much more than living expenses.

The next year we planted about twelve acres of cotton. That year the boll weevil hit South Carolina. We found the cotton squares turning yellow and falling to the ground in great numbers. We picked up squares and burned them, we poisoned the cotton; but our cotton squares continued to fall. That year we made almost no cotton and had no other money crop. It was impossible to make a payment on the farm.

The next year we again planted cotton. What else could we plant? We also planted some truck but there was no aid in marketing it, and so we lost on the produce. We put kerosene on sacks and dragged these sacks across the cotton every week or two to keep the boll weevils away. I remember very well having often found as many as twenty-four boll weevils in one cotton blossom. We were in despair. What could we do? Nothing we tried did any good.

We made no crop to speak of. The creditor offered my father the choice of share cropping or finding some money somewhere. My father was closed out. The next year we lived in poverty, selling everything we could to live on. The next year my father moved to North Carolina and at the age of fifty started life over again at a salary of $60 a month and with six children. My mother cannot bear the idea of a farm since we had so miserably failed, but father wants to try again and is looking for a little farm where he can start the game once more. I like to grow cotton, but I have little faith in it as a money crop.[31]

THE LANDLORD

The highest position in the tenure ladder is that of landlord. The passing of the old planter and the rise of the new small town merchant into absentee ownership as a speculative venture is shown in the following case.[32] A genuine love of the soil and a desire for the increased social status of the landlord can be seen:

Barnes Dickson died soon after the close of the Civil War, and his modest cotton plantation in the Arkansas River bottoms, mortgaged and involved, went into the hands of a factor. He left a son, Willey, age seven, who went to live with his elder brother, Hubert. Hubert had taken to the uplands where land was cheap and was working a "hill farm." Willey went behind the plow and stayed there until he left

[31] Abstracted from the letter of a student in a southern university, May, 1928.

[32] Summarized from an interview.

to teach school. He had never gone beyond the fourth grade, but he got the county school because he could work arithmetic, and nobody better prepared applied for the $20 a month job. From school-teaching he went to the little town of Jonesville to keep books for Josh Henry, General Merchant and Cotton Buyer, who owned a lot of farms and did a furnishing business.

Willey Dickson wanted to be a lawyer, but instead he went to a Memphis, Tennessee, business college for three months and got to be head bookkeeper at Henry's General Merchandise. When he got a chance he set up in business for himself and because he "knew niggers" he got a lot of furnishing trade. He didn't want to buy cotton, but he did want land. His first farm was a hill farm, his second was a river bottom farm bought from a Swede who had immigrated after the Civil War and, not liking cotton farming, had sold out to go West. On this last farm he put a mortgage and three Negro tenant families. The tenants later moved but the mortgage remained. In 1921 his General Merchandise failed because of bad cotton debts, but Dickson kept the farms. Since that year the farm has several times failed to pay interest and taxes. He has sold the tenants mules on credit, which they have not yet been able to pay for, and each spring he goes to the bank and borrows $500 at 10 per cent to furnish them through the year. In 1926 he bought another farm, mortgaged it, and is now furnishing three families. The 1927 flood would have ruined him had not the Red Cross taken over the burden of looking after the tenants. As it was, they made only a third of a crop. Dickson is a business man by necessity and a landowner by choice. To walk over his renters' cotton fields on Sunday afternoon is his form of nature worship. If a pretty farm were offered on the block tomorrow he would buy it if he could.

How one cotton planter was tempted to expand and how his speculations failed because of poor seasons is told in the life of Ben R. Tillman, one-time governor of and senator from South Carolina.[33] He said:

> I cleared money up to 1881 and bought land and mules right along. In that year I ran thirty plows, bought guano, rations, etc., as usual, and the devil tempted me to buy a steam engine and other machinery, amounting to two thousand dollars, all on credit. My motto was "It takes money to make money, and nothing risk nothing have." To have been entirely free from debt would have made me feel like a "kite without a tail," so I struck out boldly into deep water. Ben Jonson says:
>
> "All men are mortal
> And do have visions."
>
> I had mine and they were rose-hued. Uninterrupted success had made me a fool. I was like the "little wanton boys who swim on bladders"; but I did not know how much of a "bladder" cotton was on land impoverished of vegetable matter in dry summer.

His biographer continues:

> The investments were followed by a "dreadful drought" and fearful losses before he was aware, thrust him into the "Red Sea." An unwise optimism led him to attempt to retrieve his fortunes by additional investments. But the result was the purchase of provisions from merchants at "sickening prices," continued crop failures in 1883, 1884, and 1885, a pocketbook "like Bill Arp's when an elephant had trod upon it," and the forced sale of much of the land he had bought.

The management of a plantation is a difficult task calling for technical knowledge of cotton culture and tact in managing the human factors. The cases which follow show how, of four planters in South Georgia, two were

[33] Simkins, *The Tillman Movement in South Carolina*, pp. 51, 52.

able to adjust their farming operations to the advent of the boll weevil and the Negro migration:[34]

I am going to take here four cotton farmers whom I know personally, all of whom live in South Georgia. All of these farmers made money growing cotton up until and including the war period. The first heavy damage the boll weevil did in that section was in 1919. All of these men lost money in 1919. At this time each man found himself facing two serious problems which he had not encountered before. The boll weevil ravished the cotton crop, and there suddenly developed a shortage of Negro tenants.

Two of these farmers seemed to be able to meet the situation and overcome these difficulties, while two of them could not.

Two of these farmers lived on the farms they operated. They were brothers-in-law, living about seven miles apart. Tom Simpson operated a twenty-mule farm and John Howell ten mules on the home place. In 1925 Mr. Simpson was operating twenty-four mules and making money, while Mr. Howell had dropped down to five mule operation and was losing each year. Both of these farmers grew enough grain and hay for home consumption. Both have enough hogs each year to give them enough meat for the family use, and both have good gardens, several milk cows and a farm flock of hens. Mr. Howell usually sells a little surplus hogs, eggs, butter, and vegetables. Mr. Simpson has plenty for home consumption but does not sell any.

The chief difference between their farming methods in 1925 was that Mr. Simpson operated his entire farm with riding cultivators. Per acre in cultivation he has just one-half as many tenants as he had in 1919 and he has never had a tenant since starting operating riding cultivators to fail to

---

[34] Letter from agricultural agent of a southern railroad, May, 1928.

make enough money to "pay out" at the end of the year. This is the way he has met the shortage of labor. He has about fifteen acres of cotton planted for each mule he feeds. For each "run bill" he furnishes there are thirty acres in cotton.

Before the advent of the boll weevil Mr. Simpson fertilized his cotton at the rate of four hundred pounds of a low grade fertilizer (8-2-2) per acre. Since 1920 he has used from one thousand to one thousand two hundred pounds of a high grade fertilizer (12-4-4) per acre. He is using much thicker spacing than he did before 1920 and has discontinued planting the damp, slower lands to cotton. He is getting yields of from three-quarters of a bale to a bale of cotton per acre. These are the ways he has met the boll weevil problem.

On the other land Mr. Howell is still using the old one-horse method of cultivating cotton. He now plants about twelve acres of cotton to the plow. His tenants often fail to make enough cotton to "pay out," and he usually has one or two who are badly in debt to him to leave him each year and has a great deal of difficulty in replacing them. He is using some heavier applications of fertilizer than he did, but not more than five hundred pounds of a low grade fertilizer per acre. At first he did not change the spacing of the cotton but in the last few years he is beginning to realize that it is necessary to leave the cotton thicker. He is selecting the better drained lands for cotton, probably because he is now letting part of the farm lie idle. His yields run from one-fifth to one-half bale per acre.

The other two farmers I shall compare are James Thomas and Alec Johnson. Both of these men were so-called "absentee landlords," living in a town in South Georgia. Both of them were interested in other businesses besides farming. Mr. Johnson died in 1926. He lost heavily in his business as well as in his farming and his farms were foreclosed and sold shortly before his death. In 1919 he was operating forty-five mules, in 1925 he was operating twenty-five. Mr. Johnson was a very old man, probably around seventy. He told me he never made any change in his farming methods after the boll weevil came except that he had so much difficulty getting labor that he had to let some of his farms "lay out." He actually reduced the amount of fertilizer he used because he could not get credit. He raised some corn and hay but usually had to buy a little feed each year. His yields ran down as low as one-tenth bale per acre and never higher than one-fifth bale, although he had better cotton land than any of the others. At the time I knew him he had gotten his farms badly infested with Johnson grass from using commercial hay, and he had the most shiftless Negro tenants in the country.

James Thomas is a most unusual man. He is generally considered very lucky. Everything he touches turns to gold, and many of his friends attribute it to James Thomas' luck. As a matter of fact he is an excellent business man, with a most pronounced talent to handle labor. He was operating in 1919 ten mules. In 1926 he was operating sixty. He does not use riding cultivators for the simple reason that he never has any trouble not only in getting all the tenants he wants but in getting the most industrious Negroes in the country. He studies his tenants. Some of them he allows to plant as much as twenty-two acres of cotton to the plow and some only ten acres. The Negreos have learned that if they can manage to get on one of Mr. Thomas' farms they will have plenty of money in the fall; and in a section where there is not sufficient farm labor to tend more than three-fourths of the land which has been cultivated, he often has five or six applications from tenants for each farm. Mr. Thomas has a most perfect control over his labor and they work under his close supervision. He is using heavy applications of high grade fertilizer; is spacing the cotton thick and never plants any but his best cotton lands to cotton. He is able to get his tenants to pick up and destroy the squares which fall after the weevil has punctured them. He is also able to get surplus labor at cotton-picking time and gets his crop out quicker and in better shape. If one of his men gets behind with his crop, Thomas is always able to get his other tenants to go and help cultivate and chop the crop. Of course he pays them for this work and charges it up to the man who got the help. His yields run from three-quarters to one bale per acre, sometimes higher.

In 1924 I went through a cut of fifty acres of Mr. Thomas' cotton with a party including one of the most prominent cotton agronomists in America. This gentleman stopped in the middle of the field and said: "Mr. Thomas, with this cotton crop, if we had the 'two quarts a month law' back in Georgia, it would seem like old times."

CHAPTER X

THE COTTON CULTURE COMPLEX

THERE EXISTS a kind of natural harmony about the cotton system. Its parts fit together so perfectly as to suggest the fatalism of design. Nature's harmony of the soil, the rainfall, the frostless season, the beaming sun, and a transplanted tropic plant fit well with a transplanted tropic race, landless white farmers, and the slow but all surviving mule to supply the world's steady demand for a cheap fabric. The spinner, the cotton buyer, the landlord, the supply merchant, and the cotton farmer form an economic harmony that often benefits all except the producer, a complex whole that is so closely interconnected that no one can suggest any place at which it may be attacked except the grower; and the grower is to change the system himself, cold comfort for advice. The most heroic measures suggested to the man most bound to the system. A resolution of a recent cotton conference, for instance, reads:

> That the delegates attending this Southwide Cotton Conference held at New Orleans, Jan. 11–12, 1928, after due and careful deliberation, hereby declare their determination to no longer submit to an unfair yoke of bondage, and voice an emphatic appeal for freedom from oppression of those interests which have so long controlled the marketing and pricing of the South's great staple crop, burdening the growers with financial depression and unbearable suffering.[1]

[1] *Report of Proceedings*, pp. 41–42.

Many of the efforts of cotton growers can, from the nature of affairs, reach no higher than rhetoric.

In a very true sense the adjustment that exists to cotton cultivation will be found in the attitudes of the human factors. Accordingly, this chapter deals with traits and attitudes in southern agricultural life growing out of cotton farming. The place of the one-horse cotton farmers in southern life and the effect of cotton culture on survival are considered as elements of the cotton culture complex. The substance of this chapter is to be regarded as conclusions and inferences drawn from the whole study. Cases are offered as examples of how these attitudes operate in cotton growing. It has been suggested that the Cotton Belt may be regarded as a cultural area,[2] and the cultivation of the cotton plant regarded as a trait of material culture growing out of what Sumner and Keller call the man-land ratio, society's adjustment to its environment. There are many processes which bear a functional relation to any one element of material culture, and these processes are to be regarded as organized into sets of attitudes and modes of social behavior. In this sense we may say that attitudes surrounding cotton growing, rising from geographic and economic factors form a cotton culture complex. Although it will be seen that in the last analysis the attitudes in the cotton culture complex are to be regarded as adjustments to geographic and economic factors, it cannot be successfully maintained that they are all rational adjustments. Received

[2] This attitude may be met with commonly enough in popular writing. "In the Belt—Black, Cotton, or Bible, as you prefer," writes a flippant journalist, "cotton is Religion, Politics, Law, Economics, and Art." See Clark Wissler, "The Culture Area Concept as a Research Lead," *American Journal of Sociology,* XXXII (May, 1927), 881–91.

as social heritages they may persist as habits of thought even after they have ceased to be useful.

Another factor which must be kept in mind is that of individual variations within the cotton culture complex. It is not part of the thesis of the cotton culture complex to maintain that cotton growers are equally inclined, for instance, to attitudes of speculation or non-coöperation. It will be found, however, that with respect to many of these traits the one-horse cotton farmers, the croppers, and tenants are more nearly standardized. In analyzing the cotton culture complex we are attempting to state in terms of cultural anthropology the problem of the psychological equipment of the human factors in cotton culture.[3] In so far as the student of contemporary culture has access to economic statistics and to written expressions of attitudes, he may be regarded as on a footing equally secure with the anthropologist studying a tribe which preserves no written records.

Among the most obvious of the material culture traits associated with cotton are the food habits of its growers. It has been shown that the immense amount of man labor in planting, chopping, and picking cotton comes at times which interfere with the cultivation of other southern crops. Consequently, the family on the one-horse cotton farm has been "driven by compulsion to the most efficient of all the foodstuffs that can be made to suffice."[4] Corn is suited to the southern climate, as it is an efficient producer of cereal carbohydrates. Its dietary properties are similar to those of wheat except that, since its proteid

[3] See Clark Wissler, *Man and Culture*, pp. 47–99, for succinct statements of concepts acceptable to anthropologists.

[4] W. J. Spillman in *Farm Income and Farm Life*, ed. Sanderson, p. 194.

lacks gluten, it does not form a dough. A dietary survey conducted during the war found that the maize kernel constituted 23 per cent of the total food intake of Tennessee and Georgia mountaineers, 32.5 per cent of that of southern Negroes, but only 1.6 per cent of the diet of 72 northern families in comfortable circumstances.[5] Hogs thrive on corn and, since they complete their growth in one season, may be regarded as comparatively efficient producers of strong meat. In fact H. P. Armsby has estimated that about 24 per cent of the energy of grain is recovered for human consumption in pork as compared with about 18 per cent in milk and 3.5 per cent in beef and mutton.[6] First are eaten the glandular organs, regarded by McCollum as protective foods because of their vitamins and animo acids. The pork of higher protein content, the muscle meats such as hams and shoulders, is eaten next. The cheaper cuts of fat pork, salt cured, become the year-round staple of diet. Sorghum and sugar cane are eminently suited to the southern climate and produce, without demanding too much labor, a food of high sugar content.

Thus it comes about that the Negro cropper, the white tenant, and the small cotton farmer live upon a basic diet of salt fat pork, corn bread, and molasses. This forms the "three M diet," meat, meal, and molasses, noted by Dr. Joseph Goldberger of the United States Health Service as pellagra producing. When cotton farmers purchase food, these are the articles of diet they purchase; first, because all three are cheap, and second, because food likes and dislikes come to be matters of habit im-

[5] *Food and the War*, U. S. Food Administration, cited in E. V. McCollum, *The Newer Knowledge of Nutrition*, p. 128.

[6] "Roast Pig," *Science*, XLVI (1917), 160.

posed by culture. Exclusive reliance on this diet impairs health and economic efficiency and thus may serve to cement the cotton farmer closer to his basic diet.

The southern rural attitudes toward the field labor of women and children to a great extent grow out of the seasonal demands of cotton. The unmechanized processes of chopping and picking call for a large amount of unskilled manual labor. The time element also enters. "The limiting factor is the amount of cotton the average farm family can pick before the cotton begins to deteriorate." [7] One small mule can easily till more cotton than the average farmer can chop and pick. It is true, then, that the most successful cotton farmer is the one who can command a large amount of human labor within his own household. "It has been said with some degree of truth," writes Alexander E. Cance, "that successful farming rests on the unpaid labor of women and children." [8] Large families are an economic asset. A young cotton tenant wrote:

> A young married man single-handed can hardly rent land to farm on, as the landowner wants a man with a large family, children large enough to work so he can realize on their labor. . . . What must the young people among the renters do? They are practically denied the land to farm on until they rear enough children to gather a good-sized cotton crop; that is what the landowners want.[9]

Children thus may be said to cost the cotton farmer less and pay him more. Forced by the demands of the plant

[7] Spillman, in Sanderson, *op. cit.*, p. 194.

[8] In Sanderson, *op. cit.*, p. 77.

[9] See *The Land Question in the Southwest*, Report of Commission on Industrial Relations, X, 9262.

and his economic needs, the one-horse cotton farmer accepts the field work of his women-folks and children as a matter of course. This attitude on the part of rural families is carried into cotton mill villages. It is everywhere met in attempts to enforce compulsory school attendance.

That exclusive devotion to cotton is an attitude to be reckoned with is recognized by all agricultural extension agencies operating in the South. As a section the South has vacillated between sober realization of the plight of her cotton farmers and pride in the magnitude of her output of the fleecy staple. Cotton cultivation has become a social habit that can hardly be broken. An observer writes of the immigrant farmers in southern Oklahoma, "They have never cultivated anything but cotton, and do not want to raise anything else." [10] When forced by price failures to the cultivation of other crops, the cotton farmer is prone to return at his first opportunity to cotton. Bradford Knapp has recounted an interview with an East Texas farmer:

"I see you have a splendid field of oats," said Mr. Knapp. "Ever plant any before?"

"No. Wouldn't have planted an acre of them this time if it wasn't for this European War," replied the farmer.

"You'll get a good price for them and you won't have to buy any winter feed, will you?"

"No. But say, have you seen what cotton's doing? Nearly nine cents this morning. And I tell you if she holds at that figure very long I'm going to plow up every foot of that oat field and plant it to cotton." [11]

---

[10] Gibbons, *op. cit.*, p. 39.

[11] George McCutcheon, *The Case for Cotton*, University of South Carolina Bulletin, 1915, pp. 1 ff.

The fluctuations in price which render the cotton farmer's life uncertain have not served to check this devotion to cotton. Periods of rising prices have led to plowing up of food crops and chopping down of peach orchards to make way for cotton. On the other hand, falling prices may throw the growers deeper in debt to landlords and supply merchants, and thus force them to renewed cultivation of the cash crop. This attitude, intangible like all attitudes, is a very real thing, growing out of basic economic and geographic factors, and limiting efforts toward diversification and restriction of acreage.

Closely connected with devotion to cotton, indeed a part of it, is the speculative attitude engendered by the fluctuations in the price of cotton. A prominent cotton factor in the Eastern Belt writes:

> This attitude—a matter of degree, a degree beyond the legitimate risks of normal business—spreads itself in a thoroughgoing way and permeates the economic life of the South. Our most successful and so-called conservative business men grow up with it and are often not aware of its dangers until a crash comes. Meantime, in general, the cotton producer, lien merchant, or dealer has no other outlook, and has learned to live from year to year on the fortunes of risks over which he has absolutely no control, and upon the hazards of which he will stake his all. And when he happens to combine some other line of business with cotton, the risks he exposes himself to are in proportion.

Two cases, typical of many unwritten southern tragedies, will serve to show how the speculative element in culture goes over into individual attitudes. In the first instance the cycle of cotton prices over a generation left an attitude of speculation and expectancy that the reverses of a lifetime could not dim:

The subject, now dead, was by profession a doctor, but having lost one leg during the Civil War at Fort Sumter—at that time just out of medical college—never again practised. Heir to a large estate, he thereafter devoted himself to farming. Being a student, theory came easy and economic speculation resulted. Year by year he calculated the number of bolls to the stalk, the number of stalks to the row, the number of rows to the field; the number of pounds to the acre, dollars to the plow, and notes to the bank. And year by year the bank curtailed his labors. Though his energies never flagged, he was little by little transformed from a man of large holdings to a man of encyclopædic knowledge in all branches of the cotton industry. He could tell at a glance how many hands high a horse was or the number of pounds of fertilizer that produced a given basket of turnips. An optimist to the end, he died a few years ago an object of charity and derision.

A rising market in cotton, such as prevailed during the World War, enables cotton dealers to extend their operations to other fields and thus tends to spread the speculative risks of cotton over wider areas of business. The following case shows speculative attitudes in the dealings of a combined cotton merchant, farmer, banker, and Ford dealer:

The subject for many years was content to operate as a station buyer of cotton in a small town. The ebb and flow of luck was satisfactory and furnished a living for himself and family. Finally, in middle life, he came upon the scene of inflation caused by the World War and in a bull market, expanded with astounding favorable results. He established branch offices for his cotton firm, and then took on a small-town bank. Money flowed in. He bought farms. Cotton sold for 40 cents. He associated himself with Mr. Ford and shared the prosperity of that King of the Road. Finally, he extended his original cross-road cotton buying to merchandising at the ports as exporter, and now his influence was at the crest. The scene shifted. He entered the deflationary inter-act with large stocks of unhedged long staple cotton, for which he had paid a dollar a pound, with large bank loans based on what now looked like fictitious security, with large land holdings on which taxes and interest loomed ahead of harvests, with a Ford car ratio which his auto clientele could not absorb. He was not unhappy. He was only more active, and therefore perhaps happier. He was willing to wait for the come-back. He waited . . . waited . . . waited . . . wearily, at last . . . and then the reckoning. He was bankrupt, and his bankruptcy carried down with him a whole countryside of honest folk. If his attitude had not been what it was, he could have saved everybody, including himself; being what it was it took him twenty-four months to realize that he had lost his own half a million and much more besides. He has not gone back to his little cotton business, nor to his bank, nor to his farms, nor to his Ford agency, because he has none of these. Today he is managing to live on money that should be paid over to his creditors.

A speculative riddle confronting those independent growers who are not harassed by creditors is whether to sell on a low market or hold for higher prices. The cotton broker, although able to hedge his purchases, is in the midst of speculative attitudes to which he often succumbs. The following interesting comment was made by a Savannah, Georgia, cotton factor before the Federal Trade Commission:

> Now cotton is a torment; the cotton trade is a torment to the legal mind and the scientific mind and the logical mind, because it is neither logical nor scientific, and in many cases it is not legal.[12]

[12] *The Cotton Trade,* Senate Document 100, 1924, pp. 316–17.

The speculative nature of cotton production has carried over into the psychological equipment of the growers in another attitude, that of non-coöperation. With the American growers possessing a virtual monopoly of the cotton crop, it is held by some that coöperative acreage regulation can be used to prevent disastrous price failures. "With proper organization facilities the necessary facts as to supplies and consumptive requirements can be put in the hands of every cotton grower in the South, months before the planting season. The burden of regulating the cotton acreage would not be heavy on any farmer if coöperative action was uniformly taken each year in every cotton county of the South." [13] But as has been shown, the profits in cotton are cyclical and speculative. They come as the result of a combination of large yields and high prices. But high prices with high yields are impossible unless acreage reductions or partial crop failures occur in other parts of the Cotton Belt. To reduce one's own acreage, however, is to fail to profit to the fullest extent by the high prices. Disastrous bumper crops are followed by appeals and campaigns to reduce cotton planting. Pledges to cut acreage, called sign-ups, are secured, and optimistic publicity concerning the resulting reduction is broadcast as encouragement to the farmers. The result is that many growers, convinced of high prices, do not reduce but even increase their acreage. About thirty years ago Sidney Lanier gave us this picture of a Georgia cotton farmer reading a newspaper plea for diversification by one Clisby. It may be regarded as a not inadequate bit of psychology:

---

[13] Harvie Jordan in *Proceedings of Southwide Cotton Conference*, held New Orleans, La., Jan. 11–12, 1928, p. 28.

And presently says he: "Hit's true,
That Clisby's head is level;
Thar's one thing farmers all must do,
To keep themselves from goin' tew
Bankruptcy and the devil.

"More corn, more corn, must plant less ground
And mustn't eat what's boughten;
Next year they'll do it, reasonin's sound;
And cotton'll fetch 'bout a dollar a pound
Tharfore I'll plant all cotton."

Thus has arisen an attitude of suspicion that is the very opposite of the coöperation needed to stabilize the cotton acreage.

Cotton impinges in yet another manner upon the culture patterns of the South. The seasonal and cyclical nature of the money income not only serves to give the cotton grower a shifting standard of living, but also serves to prevent him from acquiring habits of thrift. It has been shown that the money income from cotton has ranged from $35 to $200 a bale and from $10 to $60 an acre. Without an income which can be counted upon, it is almost impossible for a family to plan and live according to a budget. The consumption of goods by the family thus tends to run in cycles corresponding to those of cotton prices rather than to be equalized over a period of years. The cotton farmer is too much given to alternate periods of splurging and deprivation.

A steady and regular flow of income, even if low, allows for saving and for planning consumption by the use of the family budget. It is thus that families of industrial wage-earners of moderate income have accepted the practice of installment buying. By saving over a period of time from unpretentious but steady wages, urban families have succeeded in providing themselves with many of the artifices of civilization. However many objections exist in regard to unwise consumption under the stimulus of installment salesmanship, one must admit that such buying can be used for wise planning of the home, partakes of the nature of the budget, and is made possible by a steady even if low income.

Such purchasing over long periods of time is not possible with the cotton farmer. He stands to make or lose his income all in one lump at one time. In the Cotton Belt luxuries are likely to be bought on the spur of the moment, during a good season in cotton, and paid for by deprivation in next year's living. Planters are apt to lament the phonographs, sewing machines, organs, player pianos, and automobiles that their tenants buy during seasons of prosperity, as evidences of inherent traits of lack of judgment and extravagance of Negroes and poor white people. The culture trait, however, draws an origin from the cyclical nature of cotton itself. A period of deprivation during the growing period is relieved by a supply of ready cash income secured virtually all at one time. Carnivals, fairs, circuses, and tent-shows realize on this fact and plan to go South during the late fall and early winter. One advertises:

PEERLESS EXPOSITION
Shows
Want Concessions

of all kinds. Now playing North Georgia where there is plenty of cotton money. Have 5 rides, 6 shows, band, 10 concessions. Positively out all winter. Haven't closed in seven years. 'Nuf said.[14]

[14] *The Billboard,* Nov., 1927.

Given as an added attitude, the fact that all cotton growers, even the planters, are bulls on cotton who find difficulty in realizing that high levels of prices are bound to fall, and one has the improvidence of the southern cotton grower. The standards of living are not only low but shifting. The income of the cotton grower has its peaks of high prices, but these peaks are not expected, they are not planned for, and they do not always serve to level up the general standard of living. Instead, they serve oftentimes to gratify whims and vagaries of consumption. The cotton farmer's income often just balances his upkeep from the landlord or a supply store. He has had nothing on which to exercise the virtue of thrift; and when his cotton crop brings him a larger money income, he is likely to regard it as a windfall and treat it accordingly.

Two other sets of attitudes peculiar to southern cotton and tobacco renters, have grown out of the conditions of tenancy rather than of cotton cultivation. The fact that they characterize an important number of cotton growers makes these attitudes an integral element of the cotton culture complex. The first attitude may be described as the shiftless attitude of the renter toward the place on which he lives. The common complaint of landlords is of houses allowed to go to ruin, fences torn down, and lands lacerated by erosion. Law gives the tenant no interest in his tenancy. A tenure of twenty years gives the renter no more right to remain than a tenure of twenty days. In this the American practice differs from the English. In addition the law gives the tenant no claim for improvements made. The tenant then does not look forward to a future but only to a present use of the farm. In self-defense his is the philosophy of get what he can while he can. To fix fences, clear land, stop gulleys from washing, to repair a shed, or shingle a roof is from his viewpoint a foolish waste of time and energy. From this attitude it may be only a step to the use of fences for fire wood. Much of the shiftlessness of southern tenants, regarded in this light, is a self-defensive adjustment.

Mobility furnishes a closely related attitude of tenants growing out of a lack of attachment to the farms which they have cultivated. In almost any region the form of share rent set by custom can take no adequate account of the variations which exist in the fertility of cotton producing soils in a given area. Variations in character of landlords also exist. Having nothing to lose, the tenant is easily led to move by a desire to secure better land, or to find a more agreeable landlord. Poor housing, inferior educational facilities and health conditions may be regarded as additional factors inciting to mobility. There is always the chance that the tenant may find a better place for no greater expenditure. Let him move a number of times, and mobility itself tends to become a habit; the renter has then acquired the reputation of being a shiftless, roving tenant. A study by the Department of Agriculture in 1922 estimated that there was a shifting of occupants on 19 per cent of all farms in the United States, 27.7 per cent of tenants and 6 per cent of owners shifting. In eight cotton states, however, 30 to 40 per cent of all farms showed a change of occupants. "White croppers reported much shorter average periods of occupancy than colored croppers," ranging from a third of a year to a year and a half.[14a]

[14a] Dept. of Agriculture Yearbook, 1923, pp. 590, 595. Kentucky, a state of tobacco tenancy, also showed a farm mobility of from 30 to 40 per cent.

When traits such as we have described are confronted in the behavior and attitudes of individuals, these traits are described in terms of personality and character defects. Two investigators close a survey of a Texas cotton growing county with the comment:

> In the editor's judgment there is one important shortcoming in the work, . . . an omission to touch with due balance upon the character of Texas tenant farmers as a cause of their own troubles. The troubles are shown, and they are shown as the troubles of the tenants. . . . Always shifting, often shiftless, sometimes unruly, the average tenant is to be regarded with pity mixed with sympathy for the one who as a business man has to deal with him.[15]

President Andrew M. Soule of Georgia State Agricultural College and L. E. Rast, agriculturist of the Arkansas Bankers' Association, have said that the inefficient, lazy farmer is one of the greatest evils with which the real farmers of the South have to contend. In a statement issued to the press, Dean D. T. Gray, University of Arkansas, College of Agriculture, said: "We want the inefficient farmers to leave and the sooner the better. What we are worried about is what will become of this type when he moves to the city." [16]

Heretofore we have proceeded more or less on the assumption that the problem of the growers of cotton is a matter of geographic and economic factors. A view commonly held is that as regards the cotton renter: "in its chief essence the problem is sociological rather than economic." The statement may be taken to mean that

---

[15] L. H. Haney and G. S. Wehrwein, *Social and Economic Survey of South Travis County,* pp. 1 ff.

[16] *Dallas News,* Feb. 17, 1928.

both biological and cultural factors are involved in the plight of the lower strata of cotton farmers. Except for those of a physical nature the biological traits of social import are difficult of appraisal. Until adequate measurements have been worked out for distinguishing and estimating inherent characteristics and until more extensive investigations are conducted, it is difficult to give factors of heredity their due. For instance, many of the traits making up the cotton culture complex may be regarded in the individual as either inherent deficiencies in character, energy, or intelligence, or as socially conditioned habits. In general the view of inherent individual differences in intelligence has come to be accepted in psychology. It is thus possible to hold that all present inefficient cotton farmers were born with inadequate hereditary equipment.

An acceptance of both the factors of inherent individual differences and their modification by culture is possible and desirable. One may rightly speak of the biological effect of the cotton farmer's environment. Nowhere in the literature is one likely to find a more eloquent presentation of the inefficient cotton grower of the South as a product of the interaction of heredity and environment than Dr. W. S. Rankin's [17] comment on the tenant, John Smith:

> Heredity started him out in life; bequeathed him his one talent; equipped him with the sort of brain and brawn that he was to use in endeavoring to . . . exercise dominion over his natural environment.
>
> But circumstance, environment, reaches back to where

[17] Formerly Secretary of the North Carolina State Board of Health, present head of the Duke Hospitalization Endowment. See *Uplift,* Oct. 22, 1921.

heredity leaves off and even before birth begins to weave silk or sackcloth for its favored or unfavored child. If unfavored, circumstance stamps the imprint of poverty upon the infant and child. . . .

Poverty draws the plan for John Smith's house; poverty tells him what he may wear; poverty decides whether he shall be warm or cold, dry or wet, clean or dirty; poverty declines to give Smith's young child its quart of milk during the day and so limits God in the sort of dust he may use in making a man; poverty denies Smith a physician to attend his wife in childbirth and sends for the old Negro midwife; poverty refuses permission to have Mrs. Smith's laceration repaired; poverty postpones sending for the physician, delays or denies needed surgery and dentistry, and acute troubles become chronic, curable passes into incurable disease.

Poverty makes Smith and his family sick. Sickness . . . keeps Smith from working or limits his output; sickness or impairment keeps Smith from thinking—thinking is the most important thing Smith should do. Smith's earning capacity is restricted and what he earns is spent on trying to recover health. Smith approves the mortgage and it makes him poorer.

Crude culture, inadequate housing, and poor health reduce energy and create inefficiency. Isolation and exclusion from educational and cultural contacts have operated to render the cotton farmer's adjustment still more inadequate. T. N. Jones, in referring to the declarations of Dean D. T. Gray and E. L. Rast, has given a vigorous statement of this view:

One fundamental aid which sustains present conditions is the uneducated condition of the masses of the people of the southern states.

The class of farmers about whom these gentlemen were talking are the product of the infamous commercial system which has existed throughout the cotton growing states since

1865. There has never been and is not now in the state of Arkansas a public free school system in which the white children, much less the Negroes, could be given any part of such training as would make them efficient in farming or anything else. . . . All of those inefficient, lazy farmers at whom the fulmination is aimed were either raised in Arkansas or some other southern state. . . .

The South to its everlasting shame, for more than sixty years has neglected the education and training of her white children except in certain favored localities, and has occupied its time and energies in building and developing urban life. Now, through some of her educators . . . she desires to disown and kick out her progeny.[18]

We have spoken of the dominating presence of national urban culture. It is possible that with respect to the lower strata of cotton producers the statement must be qualified. When an economist says that the most pathetic position in modern American economic life is that held by 40 per cent of the South's cotton farmers, the statement has implications both social and economic. There exists a group which is comparatively excluded from culture. Walter Hines Page's phrasing of the "forgotten man" senses the fact. Booker T. Washington's formula, one as the hand but separate as the fingers, was an attempt to resolve the dilemma for his own racial group.

It is known there has grown up a strongly marked division between classes in the agricultural South. C. E. Gibbons speaks of "two well defined groups of people—the banker-merchant-landlord class and the tenant-small landowner class. The former class has largely made its money from the latter class." [19] "In such areas," writes

[18] *Dallas News*, Feb. 17, 1928. [19] *Op. cit.*, p. 52.

Carl C. Taylor, ". . . as the tenant-cropper areas of the South the only semblance of a modern civilization that exists is in the country towns."[20] The presence of what he calls a maximum economic status group and a minimum economic status group, helps to account for what may be termed the rift in culture. Schools, for instance, are good in the towns, but poor in the country regions. There is a circulation of only one newspaper for every 12.7 persons in the nine leading cotton and tobacco states, compared to the rate of one paper for every 3.6 persons in the United States as a whole. The number of native white illiterates for the United States forms 2 per cent of the population. For these nine southern states it forms 5.9 per cent, and if the Negroes and foreign born are included it is 13.2 per cent.[21] The low status farmers in the South are notable sufferers from exclusion. Economic in part, their exclusion means isolation from participation in culture.

But standards of living, as E. L. Kirkpatrick[22] says, are both economic and cultural in that they are determined in part by earnings and in part by farmers' ideas of what they should buy with their earnings. This is best shown by the fact, recognized by everyone, that the cost of food may not indicate how well the members of the farm family are nourished, nor the cost of clothing how fashionably and comfortably they are dressed. Crude culture, poor taste in clothing and house furnishing, ill-chosen and ill-prepared diet, low ratios of expenditure for education, recreation, and reading are matters of contacts, training, and education. The standard of living

[20] In Sanderson, *op. cit.*, p. 149.

[21] *Loc. cit.*

[22] *Ibid.*, pp. 125–34.

is thus a culture complex. It may even serve to determine income as in the case of a cotton mill worker who makes enough to live on in four days at the mill and lays off the other two. In such a case participation in another culture by raising his desires should raise his income. The manager of a large Arkansas plantation owned by Frank O. Lowden, writes of the cotton laborer:

> He has nothing, wants nothing, expects nothing, does not try to have anything but does waste and destroy any and everything. He is wild for money, but when he gets it, it is not worth five cents on the dollar to buy his needs. That is for waste, his needs are bought on credit.

The cotton farmer is isolated from patterns of social culture by his life in the open country, by his economic status, by his lack of educational advantages, and by the doctrine of native inferiority. Thus because of social cleavage, we have another vicious circle: low economic status and the rift in culture.

If the risks of failure are great and the standards of living are low, the question may be asked, how do almost ten million southern men, women, and children manage to remain in business as cotton growers? Why do not the inefficient producers fail and pass out of the picture? The answer is that the very nature of cotton culture seems to make for the survival of inadequate farmers. The very nature of cotton tenancy in the South attracts men who otherwise could not enter farming. The tenant has no expenses of upkeep of land or buildings. He pays no taxes. He spends little or no time improving the farmstead or making it attractive. He is invited to enter the field of cotton production, is furnished a house, work stock, and a living, and becomes a first charge upon the industry.

No matter how inefficient his cultural practices, some cotton can be grown. No matter how poor the grade and texture, cotton always brings a price. The ease of cotton culture has made for the survival of inefficient men. "Any fool can grow cotton," is a saying heard all over the South. A leading citizen of North Carolina used to declare: "I wish it was harder to grow cotton. Then there would be some money in it. As it is now, any fool nigger with a bull yearlin' can make cotton." [23] The bull yearling is a thing of the past, but farmers continue to neglect the cultivation of the crop, to leave it to discolor, unpicked in the field, or unstored subject to country damage. The very methods of bagging, baling, and sampling would not be tolerated in an organized business. The southern cotton bale is the worst packaged commodity in modern commerce. It has been allowed so to remain simply because it is arranged that the damage and the tare are deducted from the sellers, the farmers, or the original shippers, and they, the least organized, can do nothing to change the practice.

In modern industry improvements in technique spread by a kind of cultural diffusion. An advance in the processing of steel soon spreads to all steel mills. It is not necessary that steel workers understand or even know of the new technique. Their need is only the "strong back and the weak mind." Technical experts attend to the installation of new processes, and executives authorize the purchase of patent rights. In agriculture every small farmer is his own executive and technical expert. David R. Coker

---

[23] J. Russell Smith, *North America,* p. 244.

may grow sea island cotton on his upland plantation at Hartsville, South Carolina; he may grow over a bale to the acre; he may conquer the boll weevil menace. True, but that is no evidence that his practices will be adopted over the Cotton Belt or even in South Carolina. Mr. Coker has said in a personal letter:

> . . . Although I and my associates have for twenty-five years been engaged in the work of improving the cotton crop, providing marketing facilities for our local territory, and trying to impress upon the South the principles of a sound and profitable agriculture, we have not been able except in a small territory to do much to check the destructive tendencies in agriculture which have for many years been undermining its foundations in large areas of the cotton states.[24]

The Department of Agriculture realizes that the farmer can have no expert guidance to correspond to the technician in industry. Consequently, thousands of dollars are spent each year in studies and experiments. The results are embodied in bulletins which are mailed to all who ask for them. But here the causal relation between ignorance and inefficient farming is clear. The vicious circle is there. Poverty created them ignorant and ignorance helps to keep them poor. Oftentimes these inefficient methods are matters of tradition, inherited from cotton growing ancestors and embedded in negative and antagonistic attitudes toward diversification, rotation, and scientific agriculture. "My folks have been cotton farmers since Hector was a pup"; "You can't tell me how to grow cotton out of a book"; "Them University profes-

---

[24] Dec. 29, 1926, in files of Rural Social Economics Library, University of North Carolina.

sors are good swivel chair farmers," are authentic expressions of folk attitudes.

In industry the inefficient processes tend to die with the failure of firms which practice them. The differential costs mean profits to the technically alert firm; to the backward concern, losses. Cotton producers, like steel producers, fail because of differential costs of production; but unlike steel mills they continue to produce and with the same inefficient methods. How can this be? If the cotton grower fails to produce a crop sufficient to pay his store bill, the bill is transferred to the next year's account. The merchant does not expect improved methods to enable his debtor to settle; instead, he hopes for the price of cotton to go up. If the tenant gets snowed under with debts, he moves and cancels them. If too many cotton farmers fail to pay up their accounts with the supply merchant, he goes bankrupt and thus shifts the load of inefficient cotton farmers back to the wholesaler. The landlord and the large planter are in a similar situation. Tenants in debt are turnips without blood. A debt hanging over a tenant may prove his greatest incentive to mobility. Rather than take the chance of securing poorer tenants, landlords sometimes forget part of the unpaid bills. Poor cotton farmers are thus protected by the semi-benevolent despotism of the cotton system. After a fashion, in good crops and in bad, they continue to survive. And with them their inefficient practices continue. They do not diversify, they are subject to all the fluctuations of the market, and they fail, but they continue to exist as cotton producers.

It must be said that to make the southern farmer merely an efficient cotton producer will not solve the problem. T. N. Jones rightly says, "There are those in this country who wish to establish a system of commercial farming." Among those economic historians who favor large scale plantation agriculture because its system of supervision increases efficiency of production are R. P. Brooks and E. M. Banks.[25] Their economic fallacy, in the writer's view, is that the plantation is fitted only to the production of cotton, and that efficient cotton production, by increasing the supply breaks the market and ruins the plantation. The sociological objection is that for farming to survive in America it must be a satisfying way of life as well as an economic adjustment. Thus it is highly possible that a somewhat inefficient diversified farmer, living under his own vine and fig tree, might make a better member of agricultural society than a highly efficient cog in a corporation plantation. The efficiency of the plantation offers but little protection against the cycles of cotton prices.

The future of the small cotton farmer is precarious at best. It may be said to hang by two threads: extension of cotton culture in the Western Belt and the possible invention of a mechanical picker. Professor R. H. Montgomery of the University of Texas predicts the virtual breakdown of cotton cultivation in the old Cotton Belt.

> West Texas and Oklahoma with their millions of acres of virgin soil and their large scale methods of cultivation can produce cotton so cheaply that the older states cannot compete with them. All that is needed to put an end to "ten acres, a nigger, and a mule" in the Old South is a mechanical cotton picker. When it is perfected—and it will be within two or three years unless all information about present experiments by the International Harvester Company and other

[25] See Banks, *Economics of Land Tenure in Georgia.*

concerns is quite incorrect—cotton production in Texas will jump unbelievably, prices will fall sharply, and the tenant farmers of the old cotton states will have to go out of business.[26]

In the meanwhile the poverty of the inefficient one-horse cotton farmer group has not operated against its survival in the succeeding generations. Exclusion from urban culture patterns has left the farmers of lower economic levels with less knowledge of contraceptive practices. For the same reasons his attitudes toward the restriction of families partake more of the old, the traditional, and the conservative. Moreover, children are less of an economic handicap in that he is not expected to do as much for them. They are more of an advantage because they furnish unskilled farm labor. After having aided in cultivating the family crop, many sons of farm owners as well of renters enter the ranks of cotton croppers and tenants. The exhaustion of free land and the differential birth rate in favor of the farmer, accounts in part for increasing tenancy rates. Cotton culture, it may be said, after rendering some of its producers inefficient, makes for their survival both in the economic and biological sense. They are the marginal farmers, for they exist on the outer margin of culture. "They constitute a more or less disturbing factor—a miserable support for themselves and a disturbing menace to the success of others that must always be counted in the estimate of production and consumption and in any proposed legislation."[27]

---

[26] Cited by Harry M. Cassidy, "The South and the Tariff," *Editorial Research Reports, 1928*, p. 599.

[27] *Condition of Agriculture in the United State*, pp. 8–9.

SELECTED BIBLIOGRAPHY

BOOKS

ARNETT, A. M.:
The Populist Movement in Georgia, Columbia University Studies, New York, 1922.

BANKS, E. M.:
Economics of Land Tenure in Georgia, Columbia University Studies, New York, 1905.

BARBEE, WILLIAM J.:
The Cotton Question, New York, 1866.

BASSETT, JOHN SPENCER:
The Plantation Overseer as Revealed in His Letters, Northampton, Mass., 1925.

BENNETT, H. H.:
Soils and Agriculture of the Southern States, New York, 1921.

BIZZELL, W. B.:
Farm Tenantry in the United States, College Station, Texas, 1921;
The Green Rising, New York, 1926.

BROOKS, E. C.:
Story of Cotton, New York, 1911.

BROOKS, ROBERT PRESTON:
The Agrarian Revolution in Georgia, 1865–1912, Madison, Wis., 1914.

BROWN, HARRY BATES:
Cotton, New York, 1927.

BRUCE, P. A.:
Plantation Negro as Freedman, New York, 1899.
Rise of the New South, Philadelphia, 1906.

BRUNHES, JEAN:
Human Geography, Chicago, 1920.

BURKETT, C. W., and POE, C. H.:
Cotton, New York, 1906.

COLLINGS, GILBERT H.:
The Production of Cotton, New York, 1926.

Condition of Agriculture in the United States and Measures for Its Improvement, Report of the Industrial Conference Board and the U. S. Chamber of Commerce, 1927.

COWPER, MARY O.:
Cotton Cloth, A Type Study in the Social Process, MS. Institute for Research in Social Science, University of North Carolina, Chapel Hill, N. C., 1925.

COX, A. B.:
The Marketing of Cotton, New York, 1929.

CRAWFORD, M. D. C.:
The Heritage of Cotton, New York, 1924.

DANA, W. B.:
Cotton from Seed to Loom, New York, 1878.

DODD, WILLIAM E.:
The Cotton Kingdom, New Haven, 1921.

DUGGAR, J. F.:
Southern Field Crops, New York, 1925.
Southern Forage Crops, New York, 1925.

EDMONDS, R. H.:
Facts about the South, Baltimore, 1907.

ENGBERG, R. G.:
Industrial Prosperity and the Farmer, Institute of Economics, New York, 1927.

EUTSLER, ROLAND B.:
Negro Agricultural Credit Conditions in North Carolina, MS. Institute for Research in Social Science, University of North Carolina, Chapel Hill, N. C., 1927.

FAULKNER, H. U.:
American Economic History, New York, 1924.

FLEMING, WALTER L.:
Civil War and Reconstruction in Alabama, Columbia University Studies, New York, 1905.
Documentary History of Reconstruction, Vols. I, II, Cleveland, 1909.
The Sequel to Appomattox, New Haven, 1921.
The Freedmen's Savings Bank, Chapel Hill, 1927.

GAINES, FRANCIS P.:
The Southern Plantation, New York, 1924.

GILBERT, W. W.:
Cotton Diseases and Their Control, New York, 1921.

GILLETT, J. M.:
Rural Sociology, New York, 1925.
GOULDING, ERNEST:
Cotton and Other Vegetable Fibers, London, 1919.
GRADY, HENRY W.:
The New South, New York, 1890.
GRUENING, ERNEST H. (ed.):
These United States, a symposium, Vols. I, II, New York, 1923–24.
HALLE, ERNST VON:
Baumwollproduktion und Pflamzungswirtschaft, Vols. I, II, Leipzig, 1897–1906.
HAMMOND, M. B.:
The Cotton Industry, Publication of American Economic Association, New York, 1897.
HART, ALBERT BUSHNELL:
The Southern South, New York, 1910.
(editor) *American History Told by Contemporaries,* 4 vols., New York, 1899–1901.
HUBBARD, W. H.:
Cotton and the Cotton Market, New York, 1927.
HUEBNER, GROVER C.:
Agricultural Commerce, New York, 1917.
HUNTINGTON, ELLSWORTH:
Principles of Human Geography, New York, 1922.
Character of Races, New Haven, 1924.
Civilization and Climate, New Haven, 1925.
Pulse of Progress, New York, 1926.
Industrial Conference Board, U. S. Chamber of Commerce. See *Condition of Agriculture in the United States and Measures for Its Improvement.*
INGLE, EDWARD:
Southern Sidelights, Baltimore, 1896.
International Cotton Buyer and Cotton Seller, Dallas, Texas, 1927.
JOHNSON, W. H.:
Cotton and Its Production, London, 1926.
JORDAN, D. S. and H. E.:
War's Aftermath, Boston, 1914.
KELLEY, WILLIAM D.:
The Old South and the New, New York, 1888.
Lea's Cotton Book, New Orleans, 1914.

McCollum, E. V.:
The Newer Knowledge of Nutrition, New York, 1922.
Mims, Edwin:
The Advancing South, New York, 1926.
Mitchell, Broadus:
Rise of Cotton Mills in the South, Baltimore, 1921.
Montgomery, R. H.:
The Coöperative Pattern in Cotton, New York, 1929.
Moore, Henry L.:
Forecasting the Yield and Price of Cotton, New York, 1917.
Mukerjee, R.:
Regional Sociology, New York, 1926.
Nourse, E. G.:
Agricultural Economics, Chicago, 1916.
Odum, Howard W. (ed.):
Southern Pioneers in Social Interpretation, Chapel Hill, N. C., 1925.
Olmsted, Frederick Law:
A Journey Through the Seaboard Slave States, New York, 1856.
A Journey in the Back Country, New York, 1860.
Otken, C. H.:
The Ills of the South, New York, 1894.
Page, Walter Hines:
Rebuilding of Old Commonwealths, New York, 1902.
Peake, R. J.:
Cotton from Raw Material to Finished Product, New York, 19—.
Phelan, John:
Readings in Rural Sociology, New York, 1920.
Phillips, Ulrich B.:
A History of Transportation in the Eastern Cotton Belt to 1860, New York, 1908.
The Plantation and Frontier, Documentary History of American Industrial Society, Vols. I, II, Cleveland, 1910.
American Negro Slavery, New York, 1918.
Potwin, Marjorie A.:
Cotton Mill People of the Piedmont, Columbia University Studies, New York, 1927.

RAMSAY, DAVID:
History of South Carolina, Charleston, 1808. 2 vols.
RHYNE, JENNINGS J.:
Case Studies in Mill Village Population, MS. Institute for Research in Social Science, University of North Carolina, Chapel Hill, N. C., 1927.
RUSSELL, ROBERT ROYAL:
Economic Aspects of Southern Sectionalism, Urbana, Ill., 1924.
SANDERSON, DWIGHT (ed.):
Farm Income and Farm Life, Chicago, 1927.
SCHERER, JAMES A. B.:
Cotton as a World Power, New York, 1916.
SCOTT, EMMETT J.:
Negro Migration During the War, New York, 1920.
SHEPPERSON, A. B. and C. W.:
Cotton Facts, New York, 1926.
SIMKINS, FRANCIS BUTLER:
The Tillman Movement in South Carolina, Durham, N. C., 1926.
SKAGGS, W. H.:
The Southern Oligarchy, New York, 1924.
SMITH, J. RUSSELL:
North America, New York, 1925.
The World's Food Resources, New York, 1919.
South in the Building of the Nation, Vols. V, VI, Richmond, 1909.
SPILLMAN, W. J.:
Balancing the Farm Output, New York, 1928.
Farm Management, New York, 1923.
TANNENBAUM, FRANK:
Darker Phases of the South, New York, 1924.
TAYLOR, HENRY C.:
Outlines of Agricultural Economics, New York, 1919.
THOMPSON, HOLLAND:
From the Cotton Field to the Cotton Mill, New York, 1906.
The New South, New Haven, 1921.
TODD, JOHN A.:
The World's Cotton Crops, London, 1924.
The Cotton World, London, 1927.

TOMPKINS, D. A.:
Cotton and Cotton Oil, Charlotte, N. C., 1901.
TURNER, F. J.:
Rise of the New West, New York, 1906.
TURPIN, EDNA:
Cotton, New York, 1924.
WARREN, G. F.:
Farm Management, New York, 1913.
WARREN, G. F., and PEARSON, F. A.:
The Agricultural Situation, New York, 1924.
WESLEY, CHARLES H.:
Negro Labor in the United States, New York, 1927.
WHITNEY, MILTON:
Soils and Civilization, New York, 1925.
WISSLER, CLARK:
Man and Culture, New York, 1923.
WOOFTER, T. J., JR.:
Negro Migration, New York, 1920.
YOUNG, T. M.:
The American Cotton Industry, New York, 1903.

PERIODICALS

Annals of the American Academy, "The Agricultural Situation in the United States," January, 1925.
ARMSBY, H. P.:
"Roast Pig," *Science,* XLVI (1917), 160.
ARNOLD, JOHN J.:
"Financing of Cotton," *Annals of the American Academy,* September, 1911.
BAKER, OLIVER E.:
"Agricultural Regions of North America, Part I," *Economic Geography*, October, 1926;
"Agricultural Regions of North America," Part II, "The South," *Economic Geography,* January, 1927.
BARROW, D. C.:
"A Georgia Plantation," *Scribner's Monthly,* April 1881.
BEATLY, B. A.:
"Negro Child Life in Rural Communities," *Proceedings of the National Conference of Social Work,* 1924.

BIZZELL, W. B.:
"Rural Housing and the Tenant Farmer," *The Survey*, April 3, 1920.

BROOKS, ROBERT PRESTON:
"The American Cotton Association," *South Atlantic Quarterly*, April, 1920.

BUTLER, TAIT:
"Farm Tenantry as It Affects Soil Fertility, *Progressive Farmer*, December 8, 1917.

CASSIDY, HARRY M.:
"The South and the Tariff," *Editorial Research Reports*, July 5, 1928 (Washington, D. C.).

COMAN, KATHERINE:
"The Negro as Peasant Farmer," *Publications of the American Statistical Association*, Vol. IX, 1904.

Commerce and Finance, Files of.

Cotton Section, *London Times Trade Supplement*, London, 1919.

COULTER, JOHN LEE:
"The Rural South," *Publications of the American Statistical Association*, XIII (1914), p. 54.

Country Gentleman, Files of.

COX, ALONZO B.:
"American Cotton in World Affairs," *Southwestern Social and Political Science Quarterly*, March, 1926, pp. 305–19.

CRAVEN, A. O.:
"The Agricultural Reformers of the Ante Bellum South," *American Historical Review*, XXXIII, 302–14.

"Dan Sully Tells the Story of His Cotton Operations," *Manufacturers' Record*, September 25, 1924.

DEEN, WILLIAM:
"The Weevil's Poison Cup," *Country Gentleman*, January 6, 1923.

DUGGAR, J. F.:
"Tenant Question as Affected by the Boll Weevil," *Progressive Farmer*, September 9, 1916.

"Economic Law and the Farm Crisis," (editorial) *New Republic*, May 6, 1925.

EDWARDS, THOMAS J.:
"The Tenant System and Some Changes Since Emancipation," *Annals of the American Academy*, XLIX, 38–46.

Fifty Years of Southern Progress, Manufacturers' Record, December 11, 1924. (Entire number.)

Fossick, G. L.:
"Ailing King Cotton," *Independent,* CXVII, 308–12.

Friday, David R.:
"The Course of Agricultural Income During the Last Twenty-Five Years," *American Economic Review,* Supplement, March, 1923, pp. 147–58.

Frissell, H. B.:
"Southern Agriculture and the Negro Farmer," *Journal American Statistical Association,* XIII, 65–70.

Geller, Carl:
"Fewer Acres and More Dollars," *Commerce and Finance,* January 12, 1927.

Gibbons, Charles E.:
"Farm Children in Oklahoma," *Child Labor Bulletin,* May, 1918.

Gist, F. W.:
"Practical Crop Diversification in the Cotton States," *Commerce and Finance,* January 12, 1927.

Grady, Henry W.:
"Cotton and Its Kingdom," *Harper's Magazine,* May, 1881.
"Era of Speculation in the South," *New York Ledger,* 1889.

Gray, Lewis:
"Southern Agriculture, Plantation System and the Negro Problem," *Annals of the American Academy,* XL, 90–99.

Gray, L. C.:
"Accumulation of Wealth by Farmers," *Proceedings of American Economic Association,* March, 1923.

Haney, Louis H.:
"Farm Credit Conditions in a Cotton State," *American Economic Review,* IV, 47–67.

Harper, Roland M.:
"Rural Standards of Living in the South," *Social Forces,* II, Nos. 1, 2, 1923.

Hibbard, B. H.:
"Tenancy in the Southern States," *Quarterly Journal of Economics,* XXVII, 482–96.

HINE, L. H.:
"Children and Cotton," *The Survey,* XXXI, 589–92.
HOLMES, G. K.:
"Peons of the South," *Annals of the American Academy,* September, 1893.
HOLSEY, A. L.:
"The Tuskegee Conference," *Social Forces,* I (1923), 287.
HOUSTON, DAVID F.:
"Cotton and the General Agricultural Outlook," *Publications of the American Economic Association,* Third Series, Vol. CXVI.
KENDRICK, BENJAMIN B.:
"Agrarian Discontent in the South, 1880–1890," *Report of the American Historical Association, 1920* (Washington, D. C.).
LEWIS, CHARLES:
"Thirty Cent Cotton and the Negro," *Illustrated World,* May, 1918.
MCKENZIE, H. B.:
"South, Cotton and the Negro," *North American Review,* April, 1924.
MCLAURIN, J. L.:
"Commercial Democracy of the South," *North American Review,* November, 1901.
Manufacturers' Record, Files of.
MILLER, E. E.:
"Cotton a National Crop," *Review of Reviews,* July, 1926.
MILTON, GEORGE FORT:
"Can Cotton Be Controlled by Law?" *Independent,* CXVII, 531–32.
MITCHELL, BROADUS:
"Fleshpots of the South," *Virginia Quarterly Review,* April, 1927.
MUNN, C. M.:
"Cotton Boll Weevil Invasion," *Independent,* February 18, 1904.
PARK, R. E.:
"The Anti-Slavery Movement in England," *American Journal of Sociology,* XXXIII (1927), 290–91.

PHILLIPS, ULRICH B.:
"Economics of the Plantation," *South Atlantic Quarterly,* July, 1903.
"Plantation as a Civilizing Factor," *Sewanee Review,* July, 1904;
"Making Cotton Pay," *World's Work,* May, 1904.
"Economic Cost of Slaveholding in the Cotton Belt," *Political Science Quarterly,* June, 1905.
"Origin and Growth of the Southern Black Belts," *American Historical Review,* July, 1906.
"Decadence of the Plantation," *Annals of the American Academy,* XXXV (1910), 37.
"Plantation with Slave Labor and Free," *American Historical Review,* Vol. XXX (July, 1925).

PICKETT, J. E.:
"World against the Cotton Grower," *Pacific Rural Press,* October 11, 1925.

POE, CLARENCE H.:
"Builders of an Agricultural Community," *South Atlantic Quarterly,* January, 1909.

PUTNAM, GEORGE E.:
"Agricultural Credit Legislation and the Tenancy Problem," *American Economic Review,* V, 805–15.

Review of Reviews, "The New South," April, 1926;
"Cotton Fields, Old and New," July, 1926, pp. 57–76.

SHEPPARD, MORRIS:
"Who Gets the Cotton Money?" *Country Gentleman,* August 23, 1924.

SMITH, ALFRED G.:
"The Cropper System," *Country Gentleman,* September 4, 1920.
"Winning the Weevil War," *Country Gentleman,* May 31, 1924.

SMITH, BRADFORD B.:
"Forecasting the Acreage of Cotton," *Journal of American Statistical Association,* March, 1925.

SNYDER, H.:
"Negro Migration and the Cotton Crop," *North American Review,* July, 1924.

SOULE, ANDREW M.:
"Problem of the Cotton Growing Industry," *Commerce and Finance,* Jan. 12, 1927.

STONE, A. H.:
"The Cotton Factorage System for Southern States," *American Historical Review*, April, 1915.
SULLY, DANIEL J.:
"King Cotton's Impoverished Retinue," *Cosmopolitan*, February, 1909.
TAYLOR, E. H.:
"Cotton Is Sitting Pretty," *Country Gentleman*, April 19, 1924.
TILDEN, FREEMAN:
"Cotton, Cotton, Who's Got the Cotton," *Country Gentleman*, January 19, 1924.
TURNER, FREDERICK JACKSON:
"Sections and Nations," *The Yale Review*, October, 1922.
TURNER, H. A.:
"The Share Renting of Farms in the United States," *International Review of Agricultural Economics*, October-December, 1923.
ULM, AARON HARDY:
"Congress Falters as to Cotton or General Farm Relief," *Commerce and Finance*, January 12, 1927.
University of North Carolina News Letter, Files of.
WANNAMAKER, J. S.:
"Stabilization of Cotton Prices," *Manufacturers' Record*, March 11, 1926.
WARNOCK, C. W.:
"The Absentee Landlord Who Demands All Cotton Rent," *Progressive Farmer*, March 29, 1919.
WARREN, G. F.:
"The Agricultural Depression," *Quarterly Journal of Economics*, February, 1924, p. 183.
WEBSTER, HENRY K.:
"Slaves of Cotton," *American Magazine*, July, 1906.
WHITE, R. CLYDE:
"Cotton and Some Aspects of Southern Civilization," *Social Forces*, September, 1924.
WHITE, WALTER:
"Negro and the Flood," *Nation*, July 22, 1927.
WILCOX, E. V.:
"The Great White Way of Cotton," *Country Gentleman*, March 31, 1922.

WISSLER, CLARK:
"The Culture Area Concept as a Research Lead," *American Journal of Sociology,* XXXII, 881–91.

REPORTS, BULLETINS, AND PAMPHLETS

AGELASTO, A. M., DOYLE, C. B., MELOY, G. S., and STINE, O. C.:
"The Cotton Situation," Dept. of Agriculture *Yearbook,* 1921, pp. 323–406.

Agricultural Crisis and Its Causes, The, Joint Commission of Agricultural Inquiry, Report 408, 67th Congress, 1st Session, 1921.

Analysis of the Management of a Cotton Growing Enterprize, Bulletin Federal Board for Vocational Education, 1926.

Atlas of American Agriculture,
Part II, *Climate,* Section A, *Precipitation and Humidity.*
Part V, *The Crops,* Section A, *Cotton,* 1918.
Part IX, *Rural Life,* Section I, *Rural Population and Organization,* 1919.

BRADLEY, F. S., and WILLIAMSON, M. A.:
Rural Children in Selected Counties of North Carolina, Children's Bureau Publication No. 33, 1918.

BRAND, CHARLES J.:
"Improved Method of Handling and Marketing Cotton," Dept. of Agriculture *Yearbook,* 1912, pp. 443–62.

BRANSON, E. C. (ed.):
Home and Farm Ownership, North Carolina Club Yearbook, Chapel Hill, N. C., 1921–22.

CARSON, W. J.:
Financing the Production and Distribution of Cotton, Federal Reserve Bulletin, 1923.

Cotton Plant, The, Dept. of Agriculture Experiment Station Bulletin 33, 1896.

Cotton Trade, The, Federal Trade Commission, Senate Document 100, 68th Congress, 1st Session, 1924.

CENSUS BUREAU, PUBLICATIONS OF:
"Cotton Production," Reports of Tenth Census, 1880, Vols. V and VI.
GOLDENWEISER, E. A., and TRUESDELL, LEON:
Farm Tenancy in the United States, 1920.
Negro Population in the United States, 1790–1918.

"Occupations," *Reports of Fourteenth Census,* IV, 1920.
Plantation Farming in the United States, Census Bulletin, 1916.
"Plantations in the South," *Reports of Thirteenth Census,* 1910, Vol. V, chap. XII.
"Southern States, The," *Reports of Fourteenth Census,* 1920, Vol. VI, *Agriculture,* Part 2.
"Southern States," *Census of Agriculture,* 1925, Part 2.
TRUESDELL, LEON:
Farm Population in the United States, U. S. Census Monograph IV.

DEPARTMENT OF AGRICULTURE, BULLETINS:
BOEGER, E. A., and GOLDENWEISER, E. A.:
Tenant System of Farming in Yazoo-Mississippi Delta, No. 337, 1916.
BRANNEN, C. O.:
Relation of Land Tenure to Plantation Organization, No. 1269, 1924.
COOK, O. F.:
One-Variety Cotton Communities, No. 1111, 1922.
Relation of Cotton Buying to Cotton Growing, No. 60, 1914.
DIXON, H. M., and HAWTHORNE, H. W.:
An Economic Study of Farming in Sumter County, Georgia, No. 492, 1917.
Dusting Cotton from Airplanes, No. 1204.
FUNK, W. C.:
Value to Farm Families of Fuel, Food, and Use of House, No. 410, 1916.
GATTIN, GEORGE O.:
Coöperative Marketing of Cotton, No. 1392, 1926.
HASKELL, E. S.:
Farm Management in Brooks County, Georgia, No. 648, 1918.
HAWTHORNE, H. W., DIXON, H. M., and MONTGOMERY, FRANK:
Farm Organization and Farm Management in Sumter County, Georgia, No. 1034, 1922.
HAWTHORNE, H. W.:
The Family Living from the Farm, No. 1338, 1925.
KIRKPATRICK, E. L., and SANDERS, J. T.:
Relation between Ability to Pay and the Standard of Living among Farmers, No. 1382, 1926.

KIRKPATRICK, E. L.:
The Farmer's Standard of Living, No. 1466, 1926.
McCONNELL, C. J.:
A Study of Cotton Market Conditions in North Carolina, No. 476, 1917.
MEADOWS, WILLIAM R.:
Sea Island Cotton Industry, No. 146, 1914.
SANDERS, J. T.:
Farm Ownership and Tenancy in the Black Prairie of Texas, No. 1068, 1922.
Studies of Primary Market Conditions in Oklahoma, No. 36.
TAYLOR, FRED:
Relation between Primary Market Prices and Qualities of Cotton, No. 457, 1916.
TURNER, H. A.:
Ownership of Tenant Farms in the United States, No. 1432, 1926.
DEPARTMENT OF AGRICULTURE, FARMERS' BULLETINS:
BAKER, OLIVER E.:
Seedtime and Harvest, Circular 183, 1922.
COOK, O. F.:
Improvements in Cotton Production, Circular 200, 1921.
CROSBY, M. A.:
An Example of Intensive Farming in the Cotton Belt, No. 519, 1913.
GILBERT, W. W.:
Cotton Diseases and Their Control, No. 1187, 1924.
GOLDENWEISER, E. A.:
The Farmer's Income, No. 746, 1916.
GOODRICH, C. L.:
Producing Family and Farm Supplies on the Cotton Farm, No. 1015, 1923.
Factors that Make for Success in Farming in the South, No. 1121, 1920.
HUNTER, W. D., and COAD, B. R.:
The Boll Weevil Problem, No. 1329, 1923.
SPILLMAN, W. J.:
Distribution of Types of Farming in the United States, No. 1289, 1923.

System of Tenant Farming and Its Results, A, No. 437, 1911.
Department of Agriculture Yearbooks, 1900–1926.
DU BOIS, W. E. B.:
Negro Landholders in Georgia, U. S. Dept. of Labor Bulletin 35.
GOLDBERGER, JOSEPH:
Pellagra in the Mississippi Flood Area, Public Health Reprint 1187, 1927.
Pellagra: Its Nature and Prevention, Public Health Reprint 1174, 1927.
GRAY, L. C., et. al.:
"Farm Ownership and Tenancy," Dept. of Agriculture *Yearbook,* 1923, pp. 507–600.
HANEY, L. H., and WEHRWEIN, G. S.:
Social and Economic Survey of South Travis County, Texas, University of Texas Bulletin 65, 1916.
Hearings before House Committee on Agriculture, December 19, 1905.
HILL, W. B.:
Rural Survey of Clarke County, Georgia, Phelps-Stokes Studies, 1915, Athens, Georgia.
HOBBS, S. H., JR.:
Know Your Own State—North Carolina, Extension Bulletin, University of North Carolina, 1925.
Land Question in the Southwest, The, Report of Commission on Industrial Relations, Senate Document 415, Vols. IX, X, 1916.
LOVALL, R. H., et al.:
Negro Migration in 1916–1917, Department of Labor, Division of Negro Economics, 1919.
MCCUTCHEON, GEORGE:
The Case for Cotton, University of South Carolina Bulletin, No. 43, Part 5, Columbia, S. C., 1915.
North Carolina Land Conditions and Problems, Report State Land Commission, 1923.
PAMPHLETS:
ATKINSON, EDWARD:
Cheap Cotton by Free Labor, New York, 1861.
Augusta Survey, Augusta, Ga., 1924.

Bailey, J. W.:
The Condition of the Farmers of North Carolina, Raleigh, N. C., 1921.
Bivens, F. J.:
The Farmer's Political Economy, Moultrie, Georgia, 1912.
Bodman, E. J.:
Foodless Farms, Leaflet. Arkansas Bankers' Association, Little Rock, Arkansas.
Gibbons, Charles E.:
See Periodicals.
Grandy, W. C. and Sons, Factors:
The Cotton Question Presented to the Cotton Trade, 1876.
Henry, Walter Richard:
Cotton and the Commission Merchants, Raleigh, N. C., 1883.
Holleman, J. T.:
Is the South in the Grip of a Cotton Oligarchy? Atlanta, Georgia, 1914.
Loring, F. C., and Atkinson, C. F.:
Cotton Culture and the South Considered, 1869.
Price, Theodore, H.:
Cotton Estates Incorporated, New York.
Proceedings of Cotton Convention Held at Memphis, Tennessee, October 13, 1926.
Proceedings of Southwide Cotton Conference, Called by American Cotton Conference, New Orleans, La., January 11–12, 1928.
Rivers, C. D.:
The Empire of Cotton, Sandersville, Georgia, 1914.
Vanderlip, Frank H.:
Our Inefficient Acres, New York, 1916.
Ransdell, Joseph E.:
Remarks in the United States Senate 1927.
Short Term Farm Credit in Texas, Texas Agricultural Experiment Station Bulletin No. 351, 1927.
Smedes, H. R.:
Agricultural Graphics 1866–1922, University of North Carolina, Extension Bulletin.
Swift, W. H.:
Child Welfare in North Carolina, National Child Labor Commission, 1918.

TAYLOR, CARL C., and ZIMMERMAN, C. C.:
Economic and Social Conditions of North Carolina Farmers, State Tenancy Commission, Raleigh, 1922.

WATKINS, D. W.:
An Investigation of Cotton Growing in Texas, Showing Influence on Cotton Production in South Carolina, Bulletin 75, Clemson, S. C.

Welfare of Children in Cotton Growing Areas of Texas, Children's Bureau Publications, No. 134, 1924.

YODER, F. R., BEARDSLEY, H. S., and HONEYCUTT, A. J.:
Farm Credit Conditions in North Carolina, North Carolina State College Bulletin, 1923.

NOTE

Much of the material used in the preparation of this study is either unpublished or not available in general libraries.

Several unpublished studies to which reference is made will be found in the files of the Institute for Research in Social Science, University of North Carolina, Chapel Hill. Mimeographed reports of the Department of Agriculture, press releases, letters, newspaper clippings, pamphlets, and rare and out of print bulletins will be found in the files of the Rural Social Economics Library, University of North Carolina, classified under the following headings:

Agricultural Production: Cotton Farm Tenancy
Marketing: Cotton
Negro in the Rural South
Rural Social Surveys
Rural Standards of Living.

Many of the case studies cited have been taken from letters from bankers, agricultural experts, general merchants, fertilizer dealers, and planters in various regions of the Cotton Belt. These letters have been deposited with the Institute for Research in Social Science, University of North Carolina, Chapel Hill, N. C.

INDEX

Fictitious names used in case studies of human factors in cotton are not indexed. Different types of cases, classified according to tenure status, may be found in the Index under *Case studies.*

www.ingramcontent.com/pod-product-compliance
Lightning Source LLC
LaVergne TN
LVHW090306210926
848480LV00001B/148